ENGLISH
RIVERS AND
CANALS

High Force Falls on the river Tees in Upper Teesdale, one of the most dramatic examples of the different ways in which rivers shape the countryside.

ENGLISH RIVERS AND CANALS

PAUL ATTERBURY

W.W. NORTON & COMPANY

NEW YORK   LONDON

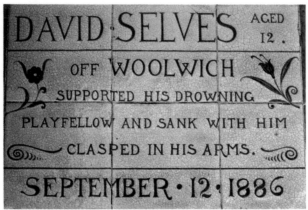

A poignant memorial at St Botolph's in the
City of London.

First American Edition, 1984

ISBN 0–393–01829–6

Filmset by Keyspools Limited, Golborne, Lancashire
Printed and bound in Italy by L.E.G.O., Vicenza

1 2 3 4 5 6 7 8 9 0

CONTENTS

The river Severn near Buildwas, Shropshire. The upper reaches of the Severn are spectacular but this stretch of the river is unnavigable.

INTRODUCTION

England is patterned with thousands of waterways that have played their part in determining its landscape and history, and one of the best ways to enjoy the English countryside is to follow their routes. The great rivers, such as the Thames, the Avon and the Severn, and those with well-established tourist appeal, notably the rivers and streams that run through the Lake District, the Yorkshire Moors, the Derbyshire Dales, the Norfolk Broads or the Costwolds, are well known. However, for every waterway that is familiar, there are literally hundreds to be discovered anew. Among these are the man-made waterways, the network of canals that revolutionised industry in the eighteenth century but which later fell into disuse.

The shape of the English landscape was established during the Ice Age, when great glaciers carved out the broad river valleys. Later, forces of erosion, particularly wind and water, cut the deeper valleys, gorges and defiles that allow rivers and streams to drain the higher ground. The nature of the English landscape is sharply varied, a pattern that is clearly evident on the ground. The most dramatic scenery is in the north, where rivers have forced their way through solid rock, cutting narrow gorges and falling precipitously into steep valleys: Teesdale in Durham and Wensleydale in Yorkshire are typical examples of this spectacular landscape. Totally different are the placid rivers of the south coast, the Arun, the Ouse and the Cuckmere, which flow gently through wide valleys in a series of meandering curves. These meanders and the oxbow lakes they often create are the legacy of far more powerful rivers that carved out the wide deep valleys in which they flow. The Thames valley, for

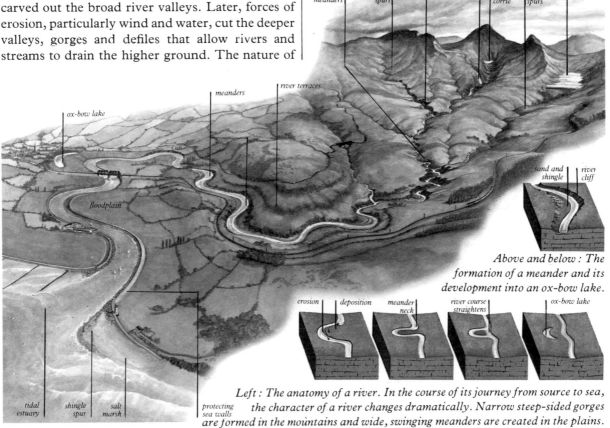

Above and below : The formation of a meander and its development into an ox-bow lake.

Left : The anatomy of a river. In the course of its journey from source to sea, the character of a river changes dramatically. Narrow steep-sided gorges are formed in the mountains and wide, swinging meanders are created in the plains.

The river Thames and the Pool of London from Tower Bridge, a view that includes some of London's most historic buildings.

example, is a wide stretch of low-lying land, flanked on the north by the Cotswolds and the Chilterns, and on the south by the Downs. Today the river, its flood plain and tributaries only fill a fraction of the huge valley that was formed in the Ice Age. From its source in the Cotswolds, the Thames falls steadily to its tidal estuary at sea level. Along its course are several large towns and cities, generally situated at important river junctions. Thus Oxford sits where the Cherwell joins the Thames, Reading surrounds the mouth of the Kennet, while London rests in a huge basin encircled by higher ground. London is situated there simply because it grew up as a settlement marking the last place where the river could be crossed before it joined the sea, and the highest point that boats could reasonably reach on the tide.

The course of a river can reflect the pattern of history over several centuries. Within a few miles its banks may reveal a Roman settlement, a medieval village, a cathedral, an eighteenth-century manor house with landscaped park, a corn mill, the wharves and warehouses of an inland port, a Victorian factory, a railway, a motorway and a modern housing estate. The river was for centuries the centre of village life, supplying water, food, power and sometimes transport, draining and irrigating the fields, supporting livestock and crops, supplying

entertainment and relaxation for the villagers and, by means of a bridge or ferry, giving the village strategic importance. A key building in any village was the mill which used water power to grind flour, the staple element in a rural diet. In the Middle Ages the mills, and thus the rivers too, were often controlled by the church, a reflection of the traditional position of the church as the source of wealth and power. Large and extravagant churches are still to be found in many riverside villages. During the gradual process of industrialisation mills and factories were built to exploit the river as a source of power, and locks and weirs were erected to make the river navigable and to allow boats to bring in raw materials for manufacturing processes and open up new markets for agricultural produce. From these small beginnings grew the Industrial Revolution, made possible by the ready availability of water power and water transport. Throughout the eighteenth century villages were turned into centres of industry, their growth encouraged by the spreading network of canals. Canal building became a mania and the most surprising and unlikely schemes were proposed. In the initial enthusiasm some of the wilder projects were actually built, plunging their backers into financial ruin. Many of these obscure canals can still be explored, picturesque memorials to misplaced optimism. Others, less obscure, are now being rescued from oblivion and restored for a new life of pleasure use.

The rivers have often played a role in the more turbulent periods of English history. The course of a waterway forms a natural barrier and for centuries boundaries have been drawn along the banks of rivers, separating countries, rival families, or regions in dispute. The routes of many rivers are marked by medieval castles built to guard a boundary or to control crossing points. Bridges, now purely objects of convenience, were originally of great strategic and economic importance, for the control of a bridge was effectively the control of the region. Today rivers still form the boundaries of a large number of counties; these may be simple administrative divisions but they are based on well-established historical precedents. Many important battles, from the Roman invasion to the Civil War, took place near rivers and their sites can be visited. Evidence of more recent military history is also hidden along the banks of the rivers and canals, notably the concrete block houses and defences hastily erected in 1940 when it appeared that every waterway would form a defensive line to hold up the German invasion.

A river is a means of draining the landscape and is thus a part of the rainwater cycle, and it is also a source of water supply. Although apparently random, this drainage pattern has been controlled by man for centuries. Certain parts of England have been radically altered by the control of water. The rich farmlands of the Fens were created from useless sodden salt marshes by the drainage schemes devised by Dutch engineers in the seventeenth century, while in the near future, the character of many major rivers could well be altered when tidal barriers turn seaways into pleasant waterways. Natural changes can also affect a river in a relatively short time. Some of the Cinque Ports of the south coast that were busy international seaports in the Middle Ages are now up to a mile inland, their harbours closed by silting.

Splendid churches in a timeless river landscape are one of the great delights of England's waterways. This fine church is at Fotheringhay, Northamptonshire.

George Cruikshank's eighteenth-century satire of a canal meeting, produced at a time when the enthusiasm for canals had become a veritable mania.

Today in England floods and droughts are rare, for water supply is controlled by a carefully balanced network of rivers, lakes and reservoirs that incorporates natural and man-made waterway courses. This network is the responsibility of a number of large regional water authorities. Apart from flood control, drainage, and domestic and industrial water supply, these authorities also deal with pollution, fishing and rights of navigation. The canals and some river navigations, such as the Weaver in Cheshire, are the responsibility of the British Waterways Board, which maintains the network, licenses pleasure boats, and looks after commercial transport and the interest of fishermen.

Increasing pollution has affected the role of the rivers during the last two centuries. Until the late eighteenth century rivers were the main source of fresh water; much of London's drinking water, for example, came from the Thames up to the early nineteenth century. The spread of industrialisation and rapid growth of towns and cities along the river banks turned the waterways into open sewers, however. Only in the last 30 years has any real progress been made in cleaning England's waterways but the strict controls now in force have brought salmon and trout back to rivers such as the Thames after a gap of over 150 years. Today the greatest problem is combating the pollution caused by farmers, whose pesticides and fertilisers are washed into small streams and make their way into larger rivers, affecting a wide range of plant and wildlife many miles away.

Fishing used to be an important part of the economic life of any tidal river but today the scale of the industry is such that most English rivers are simply too small to play a part in it. Commercial fishing survives in the oyster beds of East Anglia and the salmon rivers of the Welsh border, for example, and many small rivers have acquired a new economic significance with the spread of fish farming. However, in most areas, fishing has now become a weekend sport. Sport fishing is an increasingly popular hobby and its many practitioners can be seen lining the canal banks or enjoying the exclusive stretches of water reserved for game fishing.

The nature of boating has also undergone a radical change. Until the twentieth century the majority of boats on England's rivers and canals were involved with trade. The different characteristics of each waterway and the variety of trades plied upon them resulted in an astonishing range of strictly localised types of boat all built in a distinctive style for a specific purpose: the Humber keel, the Severn trow, the Thames barge, the Broadlands wherry, the canal narrowboat. The decline in water transport during this century has caused many of these boat types to become extinct, while others survive solely through careful preservation by enthusiasts. A trip to a river navigation or canal today will reveal a great range of boats but nearly all of these are for pleasure use; built in a universal style they are equally at home on the Ure or the Wey. Watching boats and ships is one of the greatest

A Humber keel sailing on the Stainforth and Keadby Canal in the late nineteenth century, when these boats were a familiar sight in the north-east.

Much loved by Constable, the meandering course of the river Stour in Essex is typical of the landscape that surrounds so many of England's smaller waterways.

pleasures of the waterways, whether they are the small skiffs, canoes or sailing dinghies common on the smaller rivers, or the great tankers and freighters sailing to and from the estuary ports.

Ports, harbours and docks are well worth exploring. England's waterways are full of reminders of the history of inland and coastal navigation from the Roman period onwards. Many towns and cities contain the remains of important, but often forgotten, river ports of the past. Norwich, Wisbech and King's Lynn reflect the maritime prosperity of the medieval period and the seventeenth century. Stourport, Bewdley and the other ports of the Severn reflect the Industrial Revolution of the eighteenth century. Liverpool, Manchester and the cities of the north-east look back to the wealth of the British Empire in the Victorian age. Felixstowe is a product of the modern container revolution. There are plenty of delightfully obscure little ports to be discovered too, for example, Winchelsea in Kent, or Morwellham in the West Country, fascinating ghosts of former maritime glory. Also interesting are the ports and harbours associated with military

history, notably the great naval dockyards that date back to the seventeenth century, such as Portsmouth and Chatham; the shipyards of the Tyne, the Wear, the Tees and the West Country where so many warships were built; and the small docks that grew in importance during the First and Second World Wars, for example the Channel ports and the river navigations of the south coast.

There are about 2000 miles of rivers and canals in England that are fully navigable and many hundreds of smaller waterways that can be explored by canoe or by some other type of portable boat. Navigable rivers and canals are usually well equipped with footpaths that follow the routes of former towing paths, and they can be explored just as well on foot. Smaller rivers and streams often flow through private land, however, and where this occurs walkers should make sure they keep to the official footpaths. England's waterways offer an extraordinary range of linear journeys, as the following pages reveal. All that is needed to enjoy them at first hand is a good map, plenty of time, a fresh eye, and a certain willingness to get out of the car and walk.

The rock-strewn and remote course of the East Okement river near Belstone provides a good introduction to the scenery of Dartmoor.

THE WEST COUNTRY

The landscape of the West Country has a wild and rugged quality, its age continuing to defy the smooth hand of civilisation. It is a dramatic landscape, rich in mystery, and one that reflects the region's spirit of independence. Over the centuries this spirit has survived successive invasions from the east and still retained its distinctive character. Even tourism, the greatest, and potentially the most damaging invasion of all, has not been able to alter radically the nature of the West Country.

To the west the landscape is rocky, thickly wooded in parts, and fringed by powerful cliffs and headlands. These features have determined the nature of the rivers which flow quickly through steep and twisting valleys before joining the sea in huge tidal estuaries that reach far inland. Further east, the rocks give way to hills that surround vast tracts of moorland. Bodmin Moor, Dartmoor and Exmoor are drained by numerous streams and rivers that flow away in all directions to wind their way to the sea. The south Devon estuaries retain the dramatic style of Cornwall, slightly softened by trees and the effects of years of cultivation, but in the north the coast and the estuaries become gradually wider and more generous, preparing for the transition to the rolling hills and flat marshlands of Somerset. These lands are riddled with waterways, many natural and many man-made for drainage purposes. Sedgemoor and the other moorlands of Somerset have a unique quality that has remained unaltered for centuries, which is why they have recently become the front line in the battle between agriculture and conservation.

The landscape has determined the development of the region. In Cornwall most major towns and cities have developed either on the coast, or in the sheltered corners of the river estuaries, a pattern that has not been changed by subsequent social or economic events. The traditional industries, fishing, boat-building and mining, are all dependent upon water, and the control of water. These industries encouraged the Cornish habit of self-reliance, an independence that has only been weakened in the last century, first by the coming of railways and road transport, and second by the gradual decline of the industries native to the region. For centuries the impact of mining has radically affected the landscape but with its gradual decay and the passage of time these effects have softened to become a major source of tourism. Only in the clay pits around St Austell is it still possible to imagine the appearance and feel of the West Country at the height of the Industrial Revolution.

Devon and Somerset are softer counties. Along the banks of their rivers, in their villages and towns, in their fine houses, parish churches and mills, in their rolling, well-ordered fields a more settled pattern of history is apparent. The rivers are longer, deeper and flow more gently, penetrating deep into the country and linking large inland towns and cities with the sea. There is a long history of river navigation, dating back to the sixteenth century and beyond, while the power of agriculture and industry in the eighteenth century inspired a most adventurous series of river and canal schemes, the relics of which can be explored throughout the region. One of the distinctive, and more confusing features of Devon and Somerset is the river names. There is an Otter and an Ottery, a Carey and a Cary, two Axes and an Exe, two Fromes and three Yeos, not to mention the Wolf, the Camel, the Deer, the Thrushel and the Parrett. It is all rather eccentric, and essentially English.

Lanhydrock House was built in the seventeenth century and has splendid formal and landscaped gardens. It stands overlooking the river Fowey in west Cornwall.

LYNTON

EXE

Barle

Tew

WINSE

BARNSTAPLE

APPLEDORE

BIDEFORD

GREAT TORRINGTON

Torridge

ALFARDISWORTHY

DEVON

BUDE

MARHAMCHURCH

HOLSWORTHY

Claw

OKEMENT

BRIDGERULE

Deer

Bude Canal

NORTH TAMERTON

OKEHAMPTON

EXET

Okehampton Castle

Castle Drogo

Ottery

Thrushel

Wolf

Tamar

CHAGFORD

Teign

LAUNCESTON

Lyd

MORETONHAMPSTEAD

□ Delabole Slate Quarry

POSTBRIDGE

Bovey

CORNWALL

TAVISTOCK

WIDECOMBE

TEIGNGRA

Powd

Camel

PADSTOW

BLISLAND

GUNNISLAKE

Tavistock Canal

Stover Canal

Pencarrow House

Cothele House

Dart

NEW

Buckland Abbey

CALSTOCK

ABB

Fowey

LISKEARD

BUCKFASTLEIGH

Lanhydrock House

Dartingt

Restormel Castle

ROCHE

LOSTWITHIEL

Dart Valley Steam Railway

Castle Dore

Looe Canal

Berry

Allen

PAR

Antony House

Saltram House

TOTNES

Pomer

Castle

ST AUSTELL

CHARLESTOWN

PLYMOUTH

Trewithen Gardens

Mount Edgcumbe House

Fal

Trelissick Gardens

Hayle

Godolphin House

Glendurgan Gardens

HELSTON

Helford

ENGLISH CHANNEL

The rivers of west Cornwall

Dramatic tidal estuaries, long fingers of water, their steep sides thickly wooded, exploring the rocky landscape, leading to secret corners accessible only by water; these are the features of the west Cornish rivers.

The river **Hayle** rises in the hills near Godolphin House, built between the fifteenth and the seventeenth centuries, and associated with the development of the Arabian horse in the eighteenth century. The course of the river is flanked by the remains of the Industrial Revolution, abandoned mines and quarries, the ruined engine houses and the mighty steam-powered beam engines that made the Revolution possible. The development of steam power was largely a Cornish activity, inspired by the need

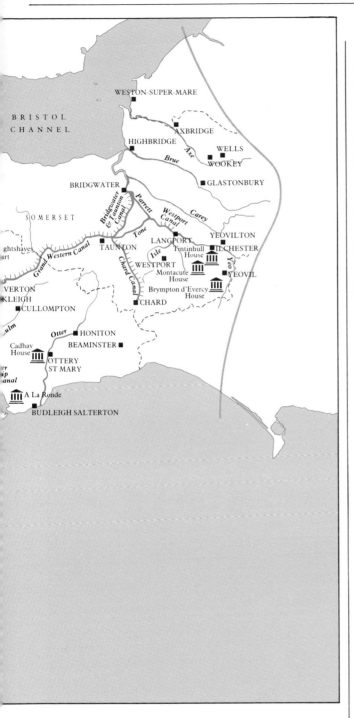

associations with Daphne du Maurier's novels. The river rises near Helston where there is a family leisure park that includes a Flambards village and an aircraft museum, while just to the north is the Wendron mining museum. The Helford estuary is known for its oysters and also for its gardens, many of which are semi-tropical because of the gulf stream. A good example is Glendurgan, at the mouth of the river. Other notable gardens, such as Trelissick and Trewithen, surround the **Fal** estuary, several miles of wandering tidal waterway linking the towns of Falmouth, Penryn and Truro. The Fal itself rises far inland, near the village of Roche with its unusual chapel, and then flows through a remote region dominated by the workings of the china clay industry before entering a wooded valley at Tregony.

The capital of the clay industry is St Austell, with its associated ports at Charleston, Par and Fowey. In the past Par was a centre for the export of the products of other industries, notably lead, tin and copper. A network of tramways and railways connected the mines with the port, and there were some small canals as well. Many of these were built by a notable mine owner, J.T.Treffry, and a suitable monument to him is a huge tramway viaduct, now isolated and forgotten in the thick woods that surround the valley of the **Par** river. To the east is the estuary of the **Fowey**, one of a number of rivers that drain Bodmin Moor. Its course is flanked by history: slate mines and subterranean caverns; the seventeenth-century splendour of Lanhydrock House with both formal and landscaped gardens; the ruins of the twelfth-century Restormel Castle; the town of Lostwithiel; while the estuary is overlooked by Castle Dore, a hillfort associated with the romantic legend of Tristan and Iseult.

Before leaving the south coast, it is worth making a slight detour to the **Looe** valley. The scenery is spectacular, a steep-sided valley cut through the rocks with at its base a fast-flowing river, a railway and a little road. At one time all methods of transport were represented here for in 1829 a canal was squeezed into the valley, linking Looe and Liskeard. Until the 1860s the canal was busy with boats full of agricultural produce, lime and, in particular, copper from the mines at Caradon Hill but the coming of the railway brought about its decline and it finally closed in 1909. Much of it has vanished, but old warehouses and the remains of some of the 25 locks can still be seen.

to drain the mines of flood waters, and Harveys of Hayle became one of the largest manufacturers of steam engines, exporting their products throughout the world from the ports of Cornwall. Having passed through the centre of Hayle, the river joins the sea in the wide sandy expanse of St Ives bay.

On the south coast is the thickly wooded estuary of the **Helford** river, noted for its seals and for its

On the north Cornish coast is the estuary of the river **Camel** which cuts its way inland to Wadebridge. The course of the river itself is a broad sweep from its source near Camelford, along the fringes of Bodmin Moor, through woodlands to the outskirts of Bodmin and then a sharp turn towards the sea. Near the river are gardens, an interesting church at Blisland, and the eighteenth-century elegance of Pencarrow House, while on the estuary are the seaside pleasures of Padstow. A tributary, the **Allen**, leads to Delabole, a slate quarry over 500 feet deep which has been worked continuously since the sixteenth century. It is the largest man-made hole in England.

The Tamar and its tributaries

The **Tamar**, which rises near Youlstone just north of Bude on the north coast, almost cuts the West Country into two and is therefore an effective boundary between Devon and Cornwall. Its upper

The waterwheel at Morwellham, which was once a busy port and is now home to an industrial museum. The quays and warehousess can still be seen.

Tavistock in south Devon. The town takes its name from the river Tavy and was linked to Morwellham by the Tavistock Canal.

reaches are remote and relatively inaccessible, to be seen only at quiet villages such as Alfardisworthy, Bridgerule and North Tamerton. It shares its seclusion with a number of tributaries that fan out in a great circle, the **Ottery**, the **Deer**, the **Claw**, the **Carey**, the **Wolf**, the **Thrushel**. These romantically named rivers can only be explored on foot. Villages are few, far between and pleasantly undeveloped, and the only town of any size is Holsworthy. At Launceston the Tamar emerges from its seclusion and then follows a winding wooded valley to Gunnislake. Along its route can be seen the achievements of both man and nature. There are many early bridges, there are fine houses, while on its tributary, the **Lyd**, there is a ruined castle, a dramatic gorge and a 90 foot waterfall. From Gunnislake southwards the Tamar is just as spectacular, but the achievements it records are of a different sort. The Tamar is tidal, and therefore navigable, from Morwellham to the sea but in the early nineteenth century boats could travel far further. The Tamar itself was extended northwards by the curiously named **Tamar Manure Navigation**, while the more significant **Tavistock Canal** was opened in 1817 to link Tavistock with

Morwellham. This featured locks, an aqueduct, inclined planes and a tunnel 2540 yards long. Morwellham became a major inland port, with wharfs able to handle 300 ton schooners. The reason for all this activity was the Devon Great Consols Mine, a group of copper mines on the east side of the Tamar that had grown up around the biggest lode of copper in Europe. They were at their peak during the first half of the nineteenth century and produced in all $3\frac{3}{4}$ million tons of copper, as well as arsenic, lead, tin and iron ore. In the late nineteenth century the industry slowly declined as the mines were exhausted and the canals closed, and gradually the whole area began to revert to its pre-industrial form. Morwellham became a ghost town, its quays overgrown, its buildings abandoned. Today, many of the remains of the Tamar's industrial boom are preserved and visitors to Morwellham can explore through them a chapter of West Country history that is hard to imagine now.

The maritime associations of the Tamar are, of course, far older, and have brought wealth to the river since the sixteenth century. The estuary is surrounded by fine houses, dating from the sixteenth to the eighteenth centuries: Buckland Abbey where Drake lived; Cothele House; Antony House; Saltram and Mount Edgcumbe. There were shipyards along the river, at Calstock and at Plymouth, a town associated with the Royal Navy since the sixteenth century, and still active as a naval base today.

Brunel's Royal Albert Bridge crosses the river Tamar at Saltash. The river is still used as a safe anchorage by large ships that moor beneath the bridge.

The Bude Canal

Although there is little to be seen today, the north Cornwall port of Bude was, in the eighteenth century, seen as the starting point for a whole network of canals that were to run for 95 miles all round north Cornwall. At the same time the promoters of the Tamar Navigation conceived a canal along the line of the Tamar to Tamerton, north of Launceston, where it would link with the Bude Canal to create a complete cross-country waterway network. In the event these grand schemes came to nought, but the idea of the Bude Canal was revived in 1817. By 1825 over thirty-five miles of canals had been built, a main line and three branches, designed primarily to carry sea sand to inland farms where it was used as a fertiliser. A feature of the **Bude Canal** were the inclined planes, whereby boats were hauled up slopes by water or steam power, rising in the process up to 200 feet. The boats themselves were fitted with iron wheels so that they could be drawn up the rails of the inclined planes without being unloaded. West Country canal engineers tended to favour inclined planes rather than locks, as typified by the Bude Canal. This extraordinary network remained in operation until the late nineteenth century when it succumbed to railway competition, but much of its route can still be discovered by the enthusiastic explorer, including the sites of some of the inclined planes, notably at Marhamchurch, Hobbacott Down and Tamerton.

This short stretch of canal and a sea lock at Bude are relics of the Bude Canal, built in the nineteenth century to carry sea sand inland for use as fertiliser.

Decorative plasterwork is a major feature of Saltram House near Plymouth, designed by Robert Adam. Saltram is easily accessible from the river Tamar.

Dartmouth in Devon. A typical West Country port set in a wooded river valley, it has been used as a safe anchorage for centuries.

Through the efforts of the canal, Holsworthy became a port town, a description still used by some today. Some of the old wharves and warehouses can be seen outside the town.

The rivers of Dartmoor

A number of rivers flow southwards from Dartmoor to the south Devon coast, of which the most important are the Dart, the Teign and the Bovey. The **Dart** is a particularly attractive river, varied in character and easy to explore by car. Its two arms rise in the centre of Dartmoor, tumbling streams crossed by early clapper bridges as they cut their way through moor and woodland. A fine place to visit is Postbridge and nearby is Widecombe, with its famous fair in September. The arms come together at Dartmeet and it flows through a wooded valley to Buckfastleigh where there are a variety of attractions including an abbey, a museum of shells and the Dart Valley Steam Railway, which is the best way to enjoy the river. To the south is the Dartington Hall arts complex and then the river reaches Totnes and the limits of the tidal estuary. A tributary leads to the romantic ruins of Berry Pomeroy Castle, buried in the woods. The Dart estuary is thickly wooded and is best seen from one of the regular boats operating between Totnes and Dartmouth. Its mouth is guarded on each side by castles.

The **Teign** is tidal from Teignmouth to Newton Abbot and then turns northwards, flanking Dartmoor until it enters a wooded valley by Dunsford. Its upper reaches are surrounded by prehistory and overlooked by Castle Drogo, one of the last country houses to be built in England. Nearby is the attractive village of Chagford.

The **Bovey** is a tributary of the Teign, branching westwards between Newton Abbot and Bovey Tracey. It rises in the heart of Dartmoor, near Moretonhampstead, and then follows a wooded course through the moor. In the 1790s a short canal was built, from Teigngrace, north of Newton Abbot to the estuary. This, the **Stover Canal**, had a remarkably useful life and remained active until 1939. Its main traffic was clay, which was shipped in vast quantities from the Teign estuary to the Mersey, from whence it was carried to the Potteries in narrow boats along the Trent and Mersey Canal. The Stover Canal was also used to transport granite from the Haytor quarries, the stone used to build, among other things, the British Museum.

The Torridge and the Taw

The **Torridge** and the **Taw** share a common estuary on the north Devon coast, a long stretch of tidal water that links Bideford and Barnstaple. Both rise in Dartmoor, the Taw directly, the Torridge via its tributaries, and both flow northwards through typical Devon scenery to the sea. The Taw is easy to explore, flowing through a variety of small villages until it reaches Barnstaple, the first town of any significance, and its route is enjoyable, though predictable. The Torridge is more varied. It is linked to Okehampton by its tributary the **Okement**, a river overlooked by the ruins of Okehampton Castle, said to be haunted by Lady Howard, a blend of the Hound of the Baskervilles and Lady Macbeth, who each night runs from Okehampton to Tavistock in the form of a hound. The upper reaches of the Torridge are quiet and inaccessible, and inspired Henry Williamson to write *Tarka the Otter*. Great Torrington, a hilltop town of style, is approached through a wooded valley. For 50 years, until the 1870s, there was a canal between Torrington and Bideford. It is mainly represented today by an impressive five-arched aqueduct over the river

Great Torrington, a handsome and traditional market town, is one of the attractions of the river Torridge in north Devon.

21

Torridge which is now used to carry the drive to Beam House, north of Torrington. Below Torrington the river becomes tidal, and sweeps through Bideford and under the medieval bridge. Bideford, a handsome town and formerly a seaport of some significance, is ranged along the west bank. From here in the seventeenth century, ships that were probably built at Appledore on the mouth of the estuary, where there is still a busy shipyard, sailed for America, loaded with pilgrims and cargoes of locally made pottery.

The Exe

Devon's major river is the **Exe**. It rises in Exmoor, south of Lynton, and then flows across the moor, through Exford, Winsford and other small villages. It is joined by a number of tributaries, the most interesting of which is the Barle whose best feature is the 17-arch prehistoric clapper bridge at Tarr Steps. Between Exbridge and Tiverton the river is in a wooded valley, overlooked by the gardens of Knightshayes Court and the medieval Tiverton Castle. Between Tiverton and Exeter the landscape is more open and the main feature is the village of Bickleigh with its working water mill, medieval castle and farm museum. Exeter is an inland port of great antiquity. There has been a canal linking the city with the Exe estuary since 1566, and since the 1830s this has been a ship canal, allowing vessels up to 400 tons to enter the city. The canal follows the course of the Exe but is quite separate from it. Its terminal basin, near the city centre, is flanked by early Victorian warehouses, markets and the Custom House of 1681. The basin now houses the Exeter Maritime Museum with its collection of boats and ships of all periods and many countries;

The wild landscape of Exmoor, which is bisected by hundreds of small, fast-flowing streams. Its rolling moorland is the haunt of wild ponies and red deer.

The Exeter Maritime Museum, which has a collection of ships and boats from all over the world, is housed in the old warehouses of the Exeter Ship Canal.

included is one of the wheeled tub boats from the Bude Canal. Originally a Roman city, Exeter has a fine cathedral and interesting buildings dating from many periods. South of the city the Exe becomes a wide waterway, overlooked by the fifteenth-century Powderham Castle and A La Ronde, a unique sixteen sided house built in 1795, before it joins the sea at Exmouth.

A major tributary of the Exe is the **Culme**. It rises in the Blackdown Hills and flows through cider country to Cullompton where it is joined by the M5 motorway, whose continuing presence spoils what was an enjoyably quiet river. Further to the east along the coast is Budleigh Salterton, and the mouth of the **Otter**. This also rises in the Blackdown Hills but its course is more interesting, passing through Honiton and Ottery St Mary, and flanked by a number of fine houses and gardens, the best of which is the sixteenth-century Cadhay House.

The Somerset rivers and canals

East of the Quantocks the landscape changes, the rolling hills giving way to the low-lying marshlands of Somerset. These are drained by innumerable ditches and streams, many of which are man-made, as well as a number of major rivers. The most significant of these is the **Parrett**, once the centre of a chain of navigable rivers and canals that were originally conceived as an inland through route from Bristol to Exmouth. This ambitious project was never completed, but the remains and relics of its surviving components make a fascinating study for those keen to understand the 'canal mania' that gripped so many speculators during the late eighteenth and early nineteenth centuries. The course of the Parrett is from its mouth at Bridgwater Bay on the Bristol Channel, a remote area of mud flats that is now a nature reserve noted for its wildfowl, through the town of Bridgwater to Langport, the present limit of navigation. Until the 1870s, it was possible for boats to continue past Langport, whose elegant Georgian waterfront reveals its former importance, and then take either the river **Isle** and the **Westport Canal** to Westport or the river **Yeo** towards Ilchester. It is still possible to find the canal basin and its associated warehouses at Westport.

The Rolle Aqueduct, a spectacular memorial to a long abandoned waterway. It used to carry the Torrington Canal over the river Torridge in north Devon.

Tarr Steps, a prehistoric clapper bridge across the river Barle in Exmoor. A simple chain of stone slabs, the clapper bridge form is primitive but effective.

Although never navigable, the upper reaches of the Yeo are particularly rich in attractions. Around Yeovil there are a number of fine houses, including Montacute, Tintinhull and Brympton D'Evercy, while at Yeovilton there is the Fleet Air Arm Museum. Nearer Langport are the ruins of Muchelney Abbey.

Another tributary of the Parrett, the **Tone**, was made navigable from 1717, linking Taunton to Bridgwater and thus to the sea. Taunton in turn became an inland port of some significance and inspired the building of a number of other canals and waterways. The first of these was the **Grand Western Canal**, an undertaking planned to link Taunton with the river Exe south of Exeter. Work began in 1810 at Tiverton, but it was not until 1838 that it reached Taunton, and then it went no further.

Within ten years it was suffering from railway competition and by 1867 the Taunton half was closed. The Tiverton end fared better and remained in use until the 1920s, and then decayed quietly until 1971 when an ambitious restoration project began to bring it back to life. Today, it is possible to cruise along the restored section of the Grand Western Canal in a horse-drawn barge. The Taunton section of the canal must have been more dramatic as it featured seven vertical boat lifts and one inclined plane, but in most cases only the sites of these strange contrivances survive.

The second canal to reach Taunton was the **Bridgwater and Taunton**, opened in 1827 to provide a more direct route between the two towns. This survived in regular use until about 1907, and is still in good condition, although isolated from other waterways. The third was the **Chard Canal**, opened in 1842 between Taunton and Chard, and closed only 25 years later. This canal featured four inclined planes, three tunnels and a few aqueducts and, considering its short life, a surprising amount can still be seen. Although unconnected to the main Somerset network, another canal in this region that should not be forgotten, the **Glastonbury Canal**, was opened in 1833, linking Highbridge on the Brue estuary with Glastonbury. This survived for only 21 years, and few remains can be seen.

The Somerset rivers and canals have much to offer the visitor, including the distinctive marshland landscape with its interesting wildlife, the many small stone-built villages, the Battle of Sedgemoor and its dramatic effects, a choice of fine houses and the remains of an ambitious waterway network that once linked towns such as Taunton, Chard and Glastonbury to the sea.

An aerial view of Montacute House taken from the north-west. Montacute is one of a number of fine houses near the upper reaches of the river Yeo.

A horse-drawn passenger barge on the restored section of the Grand Western Canal near Tiverton. A popular attraction, it is the ideal way to enjoy the Canal.

The long Albert Street cutting which leads to Bridgwater docks, on the Bridgwater and Taunton Canal in Somerset.

The Axe

The river **Axe** marks the boundary between the moors and marshlands of Somerset and the Mendip Hills that rise sharply to the north. Rising, with its tributaries near Wells, it is a river of contrasts. In Wells the cathedral with its west front embellished with nearly 400 statues, its cathedral close and its bishop's palace provides a haven of quiet in a town badly affected by traffic. To the west lies the village of Wookey, famous for the series of caves through which the Axe flows, inhabited by Stone Age and Iron Age hunters as wall as later figures of legend such as the Witch of Wookey. More recently the underground waters of the Axe have been used for making paper. The limestone forming the western flank of the Mendips is riddled with caves, many formed by rivers which now flow underground, or disappear down small holes known as 'swallets'. The river continues to Axebridge, a handsome town built round a large central square with houses of all periods including a Tudor merchant's house known as King John's Hunting Lodge. To the west it curves round Bleadon Hill and then joins the tidal estuary that leads to Weston Bay and Weston-super-Mare.

Fairground horses at Wookey Hole Museum, which boasts a number of attractions apart from its celebrated series of caves.

The old Cinque Port of Rye on the river Rother. Now nearly two miles from the sea, Rye was a medieval port of some stature.

THE SOUTH

It is the southern counties that flank the Channel which have witnessed the continuing pattern of social and economic change in England over the last 2000 years to the greatest extent. The rivers of the region have played a dominant part in the changes, and so anyone seeking to understand the history of Kent, Surrey, Sussex, Hampshire and Dorset should first look at the waterways that run through those counties.

When the Romans landed in Kent, their invasion forces were dependent upon water transport and so their beach heads were established around the mouths of the major rivers. Later, when Christianity was introduced into the country, its gradual spread followed a similar route, leaving a legacy of cathedrals, abbeys and small parish churches of a diversity probably unmatched elsewhere in England. Following the collapse of the Roman empire, waves of invading forces swept across England, all leaving their mark firmly on the southern counties. The last of these was the Norman invasion of 1066, the traces of which can be found throughout the region, in its architecture, in its place names, in the development of trade via the rivers, and in the links with France which are still very tangible today. In the early nineteenth century when the threat of invasion occurred, once more water was seen as the key to success or failure. River mouths were fortified, harbours were protected, and new waterways created to guard English ships from the threat of French guns. More recently the southern rivers have played their part again in safeguarding England's future, first by supplying the boats and the seamen that made possible the Dunkirk evacuation, and second by supporting the Normandy invasion of 1944.

Until the eighteenth century, this region had a very different appearance. In the east the wild and desolate marshlands of Kent gave way gradually to the dense oak forests that covered much of Surrey, Sussex and Hampshire, which in turn yielded to the rolling hills and plains of Dorset. During the seventeenth and eighteenth centuries most of the oak forests were felled to build ships and houses, and for the charcoal required for the smelting of iron. Both these industries were heavily dependent upon the rivers of the region. Later, as the wilder parts of Kent were brought under control, it became a centre for sheep and fruit farming and for brewing. Alongside agriculture, other industries developed, all making use of rivers and water power. During this period many of the rivers were made navigable, supplying a vital transport network that survived until the coming of the railways.

With the spread of industrialisation, the rivers became less important, reverting to their natural roles as controllers of the water supply and as adjuncts to agriculture. However, as the population of the region steadily increased, so the rivers acquired a new role, as centres of leisure and relaxation. Rivers were found to be an ideal relief from

Stained glass in Canterbury Cathedral showing pilgrims at Becket's tomb. Like many English cathedrals it has a riverside location, and stands on the river Stour.

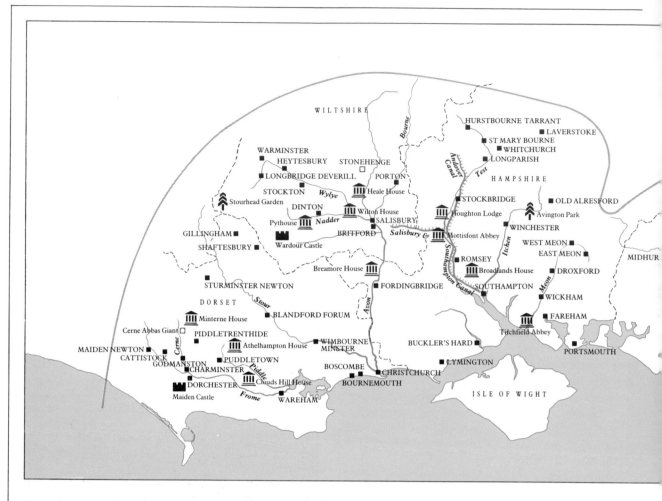

the pressures of suburban and commercial life, and the last 50 years have witnessed a great growth in activities such as fishing and boating. The rivers of southern England are becoming important again as their leisure potential is gradually realised.

The Stour

The **Stour**, which flows in a broad loop through Kent from the Downs above Folkestone to the sea at Pegwell Bay 20 or so miles along the coast, is a river that reflects much of the history of the region. Near its mouth is the great Roman fort of Richborough, one of the most exciting Roman buildings in England, and a vital defence for the Roman links between England and the European mainland. At the west gate of the fort, Watling Street started, an artery holding together the centres of Roman occupation. Nearby is the Cinque port of Sandwich, a town that is a port only in name for the continually changing course of the river has left it over a mile inland. From Sandwich the river meanders through the flat Kent marshlands to Canterbury, the tra-

ditional heart of Christianity in England whose roots were established by St Augustine in AD 597. Today the city is dominated by its cathedral, one of the four great cathedrals of southern England. Nearer its source, the Stour reveals other aspects of its past. The wealth of the area during the Jacobean period can be judged by the formal splendour of houses such as Chilham Castle and Godington Park, while the working water mill at Swanton underlines the importance of the river as a source of power prior to the Industrial Revolution. The tidal section of the river was navigable from the Middle Ages to the late nineteenth century, and for centuries Fordwich was a busy port for Canterbury, handling wool, timber and stone. However, the Stour played no part in more recent industrial activities. The railway town of Ashford straddles the river but makes nothing of it, as though expressing the dominant power of steam in the Victorian period, but nearby at Wye the river comes into its own as a source of water for the rich farmlands that have made Kent into the Garden of England.

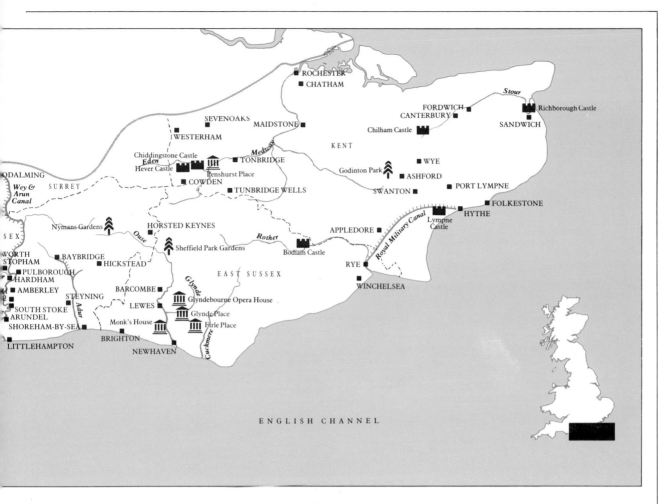

The Medway

The **Medway** is one of the forgotten rivers of England, yet it flows through attractive countryside, through towns and villages rich with history and, as a waterway navigable from Tonbridge to the Thames estuary, it has great potential for future development. It rises in the Weald and near its source at Cowden there is a furnace pond, one of the many visible reminders of the iron smelting industry of the seventeenth century. Although no longer wooded, the river retains the memory of this period until Tonbridge, a memory underlined by its tributary the **Eden** which links the great houses of Hever Castle, Chiddingstone Castle and Penshurst Place, all of which indicate the wealth of the region in Tudor and later periods. At Tonbridge the character of the Medway changes as it becomes a river navigation, one of the few survivors in this region. Locks, handsome bridges and riverside pubs now determine the nature of the river as it flows towards Maidstone, passing the oast houses at Beltring and other villages. The brewing industry originally

Lower Fittleworth Mill on the river Rother in West Sussex. The mill is one of many attractive buildings along the course of the river.

A peaceful rural scene at Teston in Kent, where a fine medieval bridge crosses the Medway, one of England's least known river navigations.

All Saints Church at Maidstone, viewed from the river Medway. The Medway remains tidal to just below Allington, where there is a sea lock.

came to Kent because of the Medway and its potential as a source of power and a means of transport. At Maidstone, the county town of Kent, the riverside is dominated by All Saints, one of the largest churches in the county, while to the north is Allington Castle, one of a number of castles built to defend the river. Others can be seen at Tonbridge and Rochester. Between Maidstone and Rochester are the obvious traces of the two great industries of the Medway, paper-making and cement, and the cathedral town of Rochester is full of interesting buildings and richly endowed with Dickensian associations. Beyond Rochester lies Chatham, until recently a naval base and garrison whose traditions go back to Henry VIII. Royal Navy ships are now no longer a regular part of the Medway scene, but there is still a great range of ships and boats to be seen, including tankers, freighters loaded with timber, pleasure boats and yachts, fishing boats and the traditional red-sailed Thames barges, preserved examples of which still sail in the Medway estuary from small harbours such as Hoo. A less visible part of the river's history is at Strood, formerly the terminus of the **Thames & Medway Canal** which ran from Gravesend to the Medway. Built to enable boats to avoid the dangerous passage round the North Foreland, the canal included in its route the second longest canal tunnel in England. The canal was closed in 1845 and few traces of it remain, except for the basin at Gravesend and the tunnel, now used by trains.

Royal Military Canal

Created between 1804 and 1806 during the invasion scares of the Napoleonic war, the **Royal Military Canal** is the only English waterway to have been built solely for defensive purposes. It formed a natural barrier against any invasion forces landing on the Romney Marshes, and allowed for the inland passage of small boats that might have been at risk at sea. The canal was later used commercially and remained in regular use until 1909. Even now it is still in reasonable condition and can be easily explored throughout much of its original length, by footpath or by minor road. The canal starts near the Cinque port of Hythe, leaving the town beside the northern terminus of the Romney, Hythe & Dymchurch Light Railway, another historical curiosity with its narrow gauge line and steam-hauled miniature trains. Surrounded by the strange scenery of the marshes, and against the backdrop of the former cliffs to the north, the canal makes its quiet way past Lympne Castle, Port Lympne with its zoo, wildlife

The Royal Military Canal. It was originally built to keep Napoleon at bay but such former glories now seem very distant.

park and gardens and the village of Appledore before joining the river Rother to the north of Rye. Another Cinque port now well inland, Rye is an interesting town that has suffered rather from tourism. Plenty of small boat activity maintains Rye's association with the sea, for the river Rother can be navigated north from Rye to Bodiam, where the fourteen-century moated castle is one of the best preserved, and most romantic examples of medieval military architecture. West of Rye the canal passes through Winchelsea before petering out near the sea at Cliff End. An exploration of this part of Kent reveals how the sea has steadily retreated over the last few centuries, leaving high and dry towns that grew up to serve the sea, including four of the five Cinque ports.

The Sussex river navigations

Along the Sussex coast, a number of rivers meet the sea in broad tidal estuaries, some of which are still active as ports. Before cutting their way through the Downs, these rivers drain the rich agricultural lands of north Sussex and Surrey, lands that in earlier centuries were thickly wooded. During the eighteenth and nineteenth centuries several of these rivers were made navigable for considerable distances through the building of locks and other engineering works. The greatest of these was the river **Ouse** which, from 1812 was navigable from the estuary at Newhaven to Upper Ryelands Bridge, 32 miles inland. Railway competition brought about the decline and final closure of the navigation in the 1860s, but a surprising amount still survives. Many

The oxbow bends of the Cuckmere River in East Sussex are a legacy of the great waterway that carved out

of the lock chambers can be found along the course of the river, while mills at Barcombe and elsewhere indicate that the river was busy in its heyday. It is worth exploring the upper reaches, to find these traces of history, to seek out the ponds that used to serve the iron smelting industry, to travel on the Bluebell steam railway at Horsted Keynes, and to enjoy the gardens at Nymans and Sheffield Park. From Lewes to the sea at Newhaven, the river was used commercially until the 1920s and still serves as an attractive link between two enjoyable, old-fashioned and lively towns. Nearby are Firle Place, and Monk's House, Rodmell, the latter the home of Virginia Woolf. A tributary, the **Glynde**, leads to Glynde Place and the Glyndebourne Opera House.

To the east is the river **Cuckmere**, a tidal river

Arundel Castle, the home of the Dukes of Norfolk for 500 years. The castle guards the gap made by the river Arun valley.

cutting its way through the Downs in a series of spectacular oxbow bends. Its upper reaches are now a centre for the English wine industry, with vineyards that can be visited. To the west lies Brighton, and beyond that Shoreham-by-Sea, a busy port on the mouth of the river **Adur**. This tidal river is still navigable for small boats through the Downs to Bines Bridge, near Steyning. During the middle of the nineteenth century the river navigation was extended to Baybridge by means of a canal with locks, the remains of which can still be found. Other features of interest along the river include Lancing College Chapel with its vast green tinted windows, and the All England Showjumping Course at Hickstead.

The most significant of the Sussex rivers is the **Arun**. It is still navigable for small boats from its estuary at Littlehampton to Arundel, a fine town clustered round the castle, the home of the Duke of Norfolk, while adventurous explorers in dinghies can follow the river northwards as far as Pallingham. Leaving Arundel the river passes the tiny villages of North and South Stoke before cutting through the Downs in a dramatic gorge at Houghton. To the east

the valley during the Ice Age.

lies Amberley, the Chalk Pits Museum of Industrial Archaeology and the fourteenth-century castle. By the village of Hardham, the Arun is joined by its tributary, the Sussex **Rother**, which flows through Midhurst, Cowdray Park and Petworth. To the north lies Pulborough, followed by the village of Stopham with its bridge of 1423, the finest medieval bridge in the region. At Pallingham, 25 miles from the sea, the navigable section ends, although the Arun continues northwards deep into the heart of Surrey. However, the Arun today is but a ghost of its former self, for in the early nineteenth century it was at the heart of a chain of navigable rivers and canals that linked London with Portsmouth and the English Channel. From the mid 1820s until 1871 boats could travel from London along the Thames to Weybridge, down the river Wey navigation to Shalford, near Guildford, along the Wey and Arun Junction Canal from Shalford to Pallingham, down the river Arun to Ford, halfway between Arundel and Littlehampton, and then to Portsmouth along the Portsmouth and Arundel Canal. This laborious and today inconceivable journey involved 116 miles of waterway, 52 locks and took at least four days. In addition, the river Rother was navigable from the Arun to Midhurst. This remarkable, and commercially unsuccessful series of waterways has been disused for over a century, and yet traces can still be found. The remains of many of the locks of the Arun and Rother navigations are still to be seen, and it is possible to follow the whole route of the **Wey and Arun Canal** on its meandering journey through some of the most attractive, and secluded, country in Sussex and Surrey. The river **Wey** is still navigable from Godalming to the Thames and so in the last ten years a Trust has been established to restore the Wey and Arun Canal to navigation. Work proceeds steadily, largely by volunteer labour, and so the dream of restoring 'London's lost route to the sea' may eventually become a reality. A century of quiet decay has been halted, but it will be many years before boats can once again travel from London to Littlehampton by the overland route.

The fishing rivers of Hampshire

As well as being the maritime heart of England, Southampton Water and its associated estuaries link the mouths of some of the prettiest, and at the same

The river Test, one of the great trout rivers of England. Here, a fly-fisherman tries his luck near Marshcourt, Hampshire.

The West Front of Winchester Cathedral. The cathedral was begun in 1079 and it is the longest in Europe, measuring 556 ft.

Buckler's Hard in Hampshire. Now a rather self-consciously preserved sailing village, it was a major ship-building centre until the eighteenth century.

time the most private rivers of southern England. To the east of the traditional ship building centres of Lymington, and Buckler's Hard on the Beaulieu river can be found the **Test**, the **Itchen** and the **Meon**. All three rivers are noted particularly for their fishing, the trout draws game fishermen to their banks from all round the world. Of these rivers the least known is probably the Meon, whose quiet valley is undisturbed by traditional riverside villages such as East Meon, West Meon and Droxford. At Wickham, a larger and more handsome town, there is a Regency watermill and a Victorian brewery. The river skirts to the west of Fareham, passing Titchfield Abbey, although by this time the peace has been shattered by motorways, railways and urban development. The Itchen is a very different river, although the countryside through which it passes is similar. At its heart lies Winchester, a city that has significantly affected the nature of the river since the twelfth century. Towards the end of that century the Bishop of Winchester, Godfrey de Lucy, built a reservoir and dredged the river to ensure an adequate water supply for his mills at Alresford and to make it navigable between Winchester and the sea. In the process he created one of the earliest river navigations in England, which survived until the fifteenth century. Restored again in 1710, the Itchen then remained in use as a navigation until the mid-nineteenth century. Exploration by car and on foot will reveal traces of several locks, and restoration of all or part of the navigation may one day be a possibility. The upper reaches are particularly attractive, with watercress beds, old mills, a steam railway, attractive but remote villages and the splendour of Avington Park, built in the Wren tradition. South of Winchester the river becomes less interesting as it approaches Southampton.

For fishermen, the most attractive river in the south of England is probably the Test, but there is far more to this river than trout. Although its banks are mostly private and therefore inaccessible, the Test is a particularly easy and rewarding river to explore by car. Rising in the Wiltshire Downs, it flows through attractive villages, Hurstbourne Tarrant, St Mary Bourne and Longparish for example, while its upper reaches are graced by two of the many unusual features that make the river so interesting. At Laverstoke is a mill that produces the paper used for banknotes, and at Whitchurch there is a silk mill. Watercress beds and thatched cottages

This sixteenth-century timber-framed fulling mill near Alresford, in Hampshire, reflects the long history of industrial use of the river Itchen.

accompany it to Stockbridge, an old coaching town. Further south the river valley is wooded and secluded, overlooked by the eighteenth-century Houghton Lodge and Mottisfont Abbey, a twelfth-century priory greatly expanded in the eighteenth century. The largest town on the Test is Romsey, dominated by its abbey, while to the south Broadlands, the former home of Lord Mountbatten, maintains the spirit of style and elegance that is characteristic of the river. Although the Test was never made navigable, its course was closely followed by the **Andover Canal**, an unrewarding venture that was in operation from 1794 to 1859. Today, little remains of the canal or its 24 locks.

The Hampshire Avon and its tributaries

The **Avon** rises in the Wiltshire Downs near Pewsey, crosses Salisbury Plain, cuts through the rolling hills of Hampshire in a wooded valley and then joins the sea at Christchurch. It is a long and remarkably varied river, matching the changing countryside through which it flows, and its salmon and trout make it attractive to fishermen. It is also a river that carries the memory of many centuries of English history, from the prehistoric mystery of Stonehenge and Woodhenge to the latest in military technology. Other centuries are represented by the Elizabethan manor house at Breamore, which also has a collection of carriages, and Heale House where King Charles was hidden for several days after the Battle of Worcester. Salisbury, with its cathedral and elegant seventeenth- and eighteenth-century architecture, reflects all the historical associations of the river. It is also a town that makes the most of its river, as can be seen in Constable's paintings. South of Salisbury is Fordingbridge, a good starting point for exploration of the New Forest, which flanks the Avon to the east. At Christchurch, a town that has spread all around the Avon estuary, there are all the pleasures of the seaside. The river Avon is unnavigable, but there were a number of attempts to link Salisbury with the sea. In the seventeenth and early eighteenth centuries the Avon was developed as a navigation and barges were able to reach Salisbury until about 1730. Apparently the Avon navigation was so fraught with difficulties that it was formally abandoned in 1772, and little trace remains today, except for an old lock near Britford. A later attempt, the **Salisbury and Southampton Canal**, which branched from the Andover Canal, was equally unsuccessful. It cost a vast amount of money to

Stonehenge is one of the better known attractions that can be discovered by exploring the remote rivers of Salisbury Plain.

build, was never completed and was only in operation for three years before being given up in 1808. Despite its short life, traces of this venture still survive but are best discovered on foot.

In spite of its lack of success in the field of navigation, Salisbury is still a waterway centre, for in and around the town the Avon is joined by a number of tributaries, the **Bourne**, the **Wylye** and the **Nadder**, all of which are worth exploring. The Bourne wanders across the Plain, through a number of small but attractive villages, as well as one or two with more sinister overtones, notably Porton and Boscombe. More interesting are the valleys of the Wylye and the Nadder. The former, which starts near Warminster, is characterised by a number of interesting churches, at Longbridge Deverill, at Heytesbury and at Stockton. The Nadder, which rises north of Shaftesbury, is a particularly well endowed river. Its valley is flanked by some fine houses, including the Palladian Pythouse, the classi-

cal style Phillips House completed in 1816, and the two Wardour Castles. One castle is the magnificent ruin of a fourteenth-century fortress embellished with Renaissance details in the sixteenth century, and the other, designed in 1768 by James Paine, is an extravagant but well-balanced mansion on a grand scale. Further to the east, by the junction of the Nadder and the Wylye, lies Wilton, traditionally the home of carpet making, and dominated by the seventeenth-century splendour of Wilton House, designed by Inigo Jones. Another feature of the Nadder is the distinctive cream coloured tufa stone used for the cottages in villages such as Dinton. This stone comes from the quarries at Chilmark, a mile to the north.

The Dorset Stour

The Dorset **Stour** has little in common with its Kentish namesake. It rises at Stourhead, one of the earliest of the eighteenth-century landscape gar-

Wilton House, home of the Earls of Pembroke for over four hundred years and one of the most splendid country houses in England.

dens, whose lakes, temples and wooded vistas anticipated the vogue for 'natural' gardens and landscapes that was to dominate the second half of the century. It flows south through Gillingham and the rolling countryside that characterises the region. By the time it reaches Sturminster Newton, a handsome and unspoilt town, the Stour has developed associations with Thomas Hardy, and in particular *Tess of the d'Urbervilles*, that are interwoven with the landscape, the villages and above all the rivers of the area. Fine towns and elegant buildings are a feature of the river, Blandford Forum, Crawford Bridge, Kingston Lacy and Wimborne Minster, as it wanders on its way to join the sea near Christchurch.

The Frome and its tributaries

The rivers of south Dorset are small and winding, but they link together many features of the county's past. Prehistory is represented by Maiden Castle, near the **Frome**, and the Cerne Abbas Giant, carved into the hills above the **Cerne**. There are abbeys, dark and romantic, such as the ruins of Bindon beside the Frome, and fine houses and gardens, notably Athelhampton on the **Piddle** and Minterne on the Cerne. There are churches, at Cattistock on the Frome, at Cerne Abbas and

Memories of Judge Jeffreys and the Monmouth Rising are never far away in Dorset. This sign hangs outside the Judge's lodgings in Dorchester.

Stourhead in Wiltshire is one of England's premier landscape gardens and it reflects the importance of water in eighteenth-century garden design.

Charminster on the Cerne and sad memorials, in Dorchester and elsewhere, of Judge Jeffrey's reign of terror following the Monmouth Rebellion. More recent history is reflected by the Tank Corps Museum at Bovington, overlooking the Frome valley. The Frome rises in the hills east of Beaminster, then flows in a steep valley through Maiden Newton to Dorchester and thence through more open country to the sea at Wareham. Its tributaries include the Cerne, which can claim, at Godmanstone, one of the smallest pubs in England, and the curiously named Piddle or Trent, with its string of even more curiously named villages: Piddletrenthide, Puddletown, Affpuddle and Briantspuddle. There is also Tolpuddle, whose fame stretches far beyond Dorset.

These rivers have strong literary connections. The spirit of Hardy is everywhere, from his birthplace at Higher Bockhampton above the Frome to the many settings that can readily be identified in his books. Less well-known is Dorset's association with T.E.Lawrence, 'Lawrence of Arabia'. His house, Cloud's Hill, on the hills above the Frome, can be visited, while in the fine Saxon church of St Martin's on the Wall at Wareham there is a fine monument, modelled by Eric Kennington.

The gentle flight of locks at Napton on the Oxford Canal show this rural English waterway at its best. At the summit is the attractive village of Napton-on-the-Hill.

THE THAMES

The Thames must be one of the most famous rivers in Europe. From its source near Cirencester to its mouth near Southend, it has been described, documented and dissected in all its aspects, and its virtues have been extolled by writers as varied as Charles Dickens, Jerome K Jerome and Kenneth Grahame. Its role in English history is so much a part of the education of successive generations of children that knowledge about the Thames is now almost instinctive. Since the invention of tourism itself, the Thames and its connections have been explored by an unceasing tide of visitors, on foot or horseback, by boat, bicycle or car. Despite all this, there are still many facets of this much-loved river that are not well known, and which warrant fresh voyages of discovery.

The Thames is a well connected river. It is blessed throughout its course by a great variety of tributaries and it has been linked by other canals and waterways to almost every corner of England. In the last 100 years a number of these links have been broken, but enough generally remains to make exploration enjoyable and worthwhile. The Thames is essentially a band drawn across England. It rises, with a number of its tributaries, in the Cotswolds which it drains to fertilize the rolling farmlands of Oxfordshire. It carves its way between the Chilterns and the Berkshire Downs, forming a wide valley

The river Thames winds its way through the heart of London, from Westminster Palace in the west to Tower Bridge and the docks in the east.

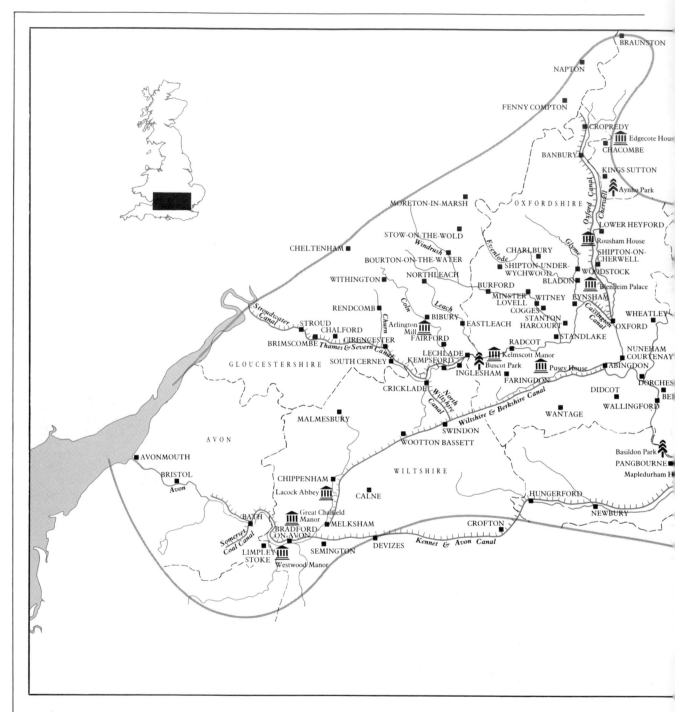

that leads it to the great basin in which London sits. From the high ground all around the basin further tributaries flow into the Thames, swelling its waters and driving them faster towards the estuary, flanked by the low-lying lands of Essex and north Kent. At its mouth the Thames is more than a mile wide. Over the centuries the river has been at the heart of English life, variously a valuable fishing river; a defensible boundary and the scene of many battles and campaigns; a transport artery bringing wealth to the towns and villages along its banks; a creator and supporter of industry; a source of fresh water; a sewer; the instigator of pageantry; and the background for leisure and relaxation.

The Thames
The **Thames** flows across the country from its source near Cirencester to its estuary by Southend,

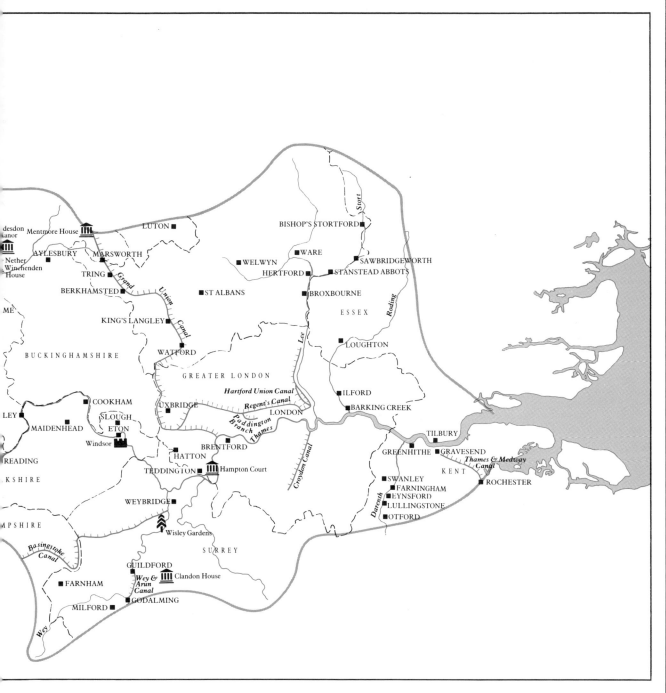

passing through towns and villages that have witnessed the making of English history. It has always been associated with boats, probably more so than any other river in England. The river has been navigable for much of its length since prehistory and today its smart locks with their decorative flower beds are a characteristic feature of one of England's most popular holiday regions. The heyday of the river was probably during the late nineteenth cen-

tury when it was used by a great variety of boats. The upper reaches then were crowded with punts, camping skiffs, sailing dinghies and a surprising range of commercial canal barges. From Oxford to Teddington there were pleasure steamers, yachts, tugs and barges, while the tideway through London to the sea was always packed with fishing boats, sailing barges, coasters and larger merchant vessels. Today, exploring the Thames by boat is still a major holiday activity, although much of the river's commercial traffic has disappeared. The stretch from Teddington to the limit of navigation at Inglesham, just west of Lechlade, is always busy with pleasure boats of all sizes, a continuous stream of traffic

The pleasures of boating on the Thames, in its Edwardian heyday, graphically illustrated in Edward Gregory's Boulter's Lock–Sunday Afternoon.

making its way through the towns and villages that give the river its broad appeal.

The Thames can for convenience be divided into sections, each of which has a distinct character and contains specific features of interest. The least known section is the unnavigable stretch from the source to Inglesham, where the attractions include the source itself, marked by a statue in a field and strangely dry for most of the year; the Cotswold Water Sports Park near South Cerney; the little town of Cricklade with its eighteenth-century houses and fine Tudor church tower; Kempsford, once the home of the Plantagenets and Fairford, where some of the best early stained glass in

England is to be seen. Inglesham church is a treasure, a tiny thirteenth-century building complete with box pews. The river in this stretch is shallow, twisting and often inaccessible, but near Inglesham the towing path begins which then continues all the way to London.

The next stretch, from Lechlade to Oxford, is particularly attractive. The broad river valley is flanked on the south by the Berkshire Downs, the Ridgeway and the Vale of White Horse, and on the north by the Cotswolds. There are many interesting towns and villages; Burford, Witney, Faringdon, Stanton Harcourt; some fine houses and gardens; Buscot, Kelmscott and Pusey; and at Radcot and Newbridge, two of the oldest bridges on the river.

Between Oxford and Reading the Thames comes into its own as a major tourist river, easily explored both by boat and by car. The best feature of this stretch is the dramatic, steeply wooded valley cut through the Chilterns, but there are many other attractions. It is an area rich in archaeology and history. The Ridgeway crosses the Thames at Streatley. There are abbeys at Abingdon, Dorchester and Reading, a castle mound at Wallingford and fine houses at Basildon and Mapledurham. The nautical college at Pangbourne is worth a visit, as is the famous rose nursery at Nuneham Courtenay and there are some unusual museums, including the steam railway centre at Didcot and a vintage bicycle collection near Benson. There is even a vineyard near Reading.

From Reading to London the Thames pursues a familiar course, passing through the famous boating centres of Henley, Cookham and Maidenhead, and following a winding course along the wooded north side of the valley. This stretch reflects many aspects of English history, most notably at Eton, Windsor and Hampton Court. After passing through the elegant and decorative outer suburbs of western London, the Thames reaches Teddington, and the tideway begins.

The London tideway used to be a great commercial river, the root of the city's prosperity, but over the last 20 years commercial traffic has steadily declined as, one by one, the London docks have closed. Today London's river has a new future as a centre for leisure activities. East of London, all the former docklands are now facing redevelopment and so the nature of the river is bound to change. Even the tides can be controlled by the new Thames barrier at Woolwich. Tilbury in Essex is now the

Henley, a decorative riverside town that makes the most of the Thames. It really comes to life when the annual rowing regatta is in progress.

centre of commercial traffic, while many ships that would formerly have docked in the heart of London now sail no further than Felixstowe in Essex, the fastest growing dock complex in Britain. The wide tidal river east of London is predominantly industrial, but it is full of interest. Traces of England's maritime past can be found at Tilbury, Greenhithe and Gravesend while further east, on the Kent shore, is a wild and little known area of marshland that is full of wildlife, old castles and defence systems, and remote villages unchanged since the time of Dickens. The Thames joins the sea flanked by oil refineries, holiday camps and caravan parks and the seaside pleasures of Southend whose mile-long pier marks the effective end of the river.

The very familiarity of the Thames may limit its appeal for many people but there is far more to the river than a simple linear pleasure park. The greatest, and at the same time the least known aspect of the Thames is its role as the centre of a major network of rivers and waterways that linked many parts of southern England with the Midlands. Today this network is considerably smaller than it was in its heyday in the early nineteenth century, but it is still extensive and well worth exploring. Like the Thames itself, its connecting network of rivers and canals can be divided into a series of sections individually attractive to visitors.

The Cotswold rivers

The Cotswolds, a landscape of high, rolling hills, open grazing land and attractive stone villages, is drained by several rivers which flow rapidly down to join the Thames. These rivers are a natural way to explore the Cotswolds for they link many of the most interesting villages. First among these rivers is the **Churn**, whose seven springs were formerly considered a rival source for the Thames. Near Rendcombe, on the Churn, is one of the surviving habitats of the large edible snail introduced by the Romans. The **Coln**, rising east of Cheltenham, is a fast-flowing trout stream that connects remote villages such as Withington with the rich architecture of Bibury, Arlington Mill and Fairford, before joining the Thames near Inglesham. The **Leach**, a secretive river that flows underground for some of its course, rises near the handsome wool town of Northleach, the home of the Cotswold Countryside Museum, passes through the village of Eastleach with its two churches and clapper bridge and then joins the Thames near Kelmscott Manor. More dramatic is the **Windrush**, the most typical of the Cotswold rivers. This rises near Stow-on-the-Wold then winds slowly along a magnificent valley to link

A glimpse of a typical Cotswold cottage at Minster Lovell on the river Windrush, which winds through some of the most attractive Cotswold country.

some of the most attractive of the Cotswold towns and villages. These include Bourton-on-the-Water with its many bridges, model village and perfumery; Burford; Minster Lovell, where the romantic ruins of the hall overlook the river; Witney, the home of the blanket industry; and Cogges with its farm museum. Before joining the Thames near Standlake, the Windrush flows through an area of gravel

The model village at Bourton-on-the-Water, crossed by many bridges, echoes the attractions of its orginal and captures the spirit of the Cotswolds in miniature.

pits, now used for water sports. The last of the Cotswold rivers is the **Evenlode**, which rises south of Moreton-in-Marsh. Remote, unchanged villages and fine churches mark its course to Shipton-under-Wychwood. Wychwood forest, now 1500 acres of natural woodland and formerly one of the greatest of English forests is flanked by the Evenlode as it passes through Charlbury. A tributary, the **Glyme**, leads to Bladon, Woodstock and Blenheim Palace, where it was widened in the eighteenth century to form the great lake. The Evenlode joins the Thames near Cassington, the site of the **Cassington Canal**, cut to link Eynsham with the Thames and long disused. None of the Cotswold rivers are navigable, but the Windrush and the Evenlode can be explored by canoe, particularly the lower reaches.

The river Cherwell and the Oxford Canal

Rising above Charwelton in Northamptonshire, the **Cherwell** links the dark stone villages of the Midlands with the honey-coloured stone of the Cotswolds and Oxfordshire. It is a placid river flowing through watermeadows and flanked by pollarded willows, but its banks are adorned with many architectural treasures, including Edgecote House, Aynho Park, and Rousham, with its early eighteenth-century landscape garden by William Kent. There are fine churches at Chacombe, Kings Sutton and Shipton-on-Cherwell. Among its towns and villages, Banbury, Cropredy and Lower Heyford are interesting. The Cherwell's approach to Oxford is particularly attractive, dominated by the tower of Magdalen College and the eighteenth-century bridge. In Oxford the Cherwell is much used for punting.

From Cropredy southwards to Oxford the Cherwell runs closely beside the **Oxford Canal** and in a number of places river and canal merge. However, the Oxford Canal is a quite distinct waterway, not simply a canalisation of the Cherwell. Opened throughout in 1790 from Oxford to Hawkesbury, near Coventry, the Oxford Canal links the Thames with the Midlands, and was conceived as part of a grand eighteenth-century scheme to link the Thames, Severn, Trent and Mersey. Until 1805 the Oxford Canal was the only route linking London with Birmingham, Liverpool, Manchester, and the other industrial cities of the north. The main traffic on the canal was coal, a trade that continued until the 1950s. Now the Oxford Canal is entirely used by pleasure boats, and is one of the most popular cruising routes. It is above all a rural canal, its winding course following the contours of the landscape. The most notable engineering features are the flight of locks up to the summit at Napton and the long cutting at Fenny Compton. A variety of

The magnificent Gothic tower of Magdalen College in Oxford dominates the river Cherwell at Magdalen Bridge, where it joins the Thames.

stone and brick arched bridges, wooded locks, small aqueducts and characteristic wooden lift bridges, plus the visual attraction of the Cherwell valley make the Oxford Canal a pleasure to explore.

The Thame

Rising near Aylesbury, the **Thame** is one of the lesser-known of the Thames tributaries. Its source is surrounded by grand houses, ranging from the Tudor elegance of Nether Winchendon to the Victorian splendour of Waddesdon, but the Thame is predominantly a quiet rural river. The main town, from which it takes its name, is Thame, an attractive but strangely little-known place with a range of interesting buildings flanking a long main street. Near Wheatley is the Waterperry Horticultural Centre, after which the river makes its way through a rather remote landscape to Dorchester and its junction with the Thames.

The Thames and Severn Canal

One of the least known features of the Thames today is its former connections with the rivers of the west of England, and in particular with the Severn. At one time, there were three cross-country canal routes between the two rivers but none of these are navigable today. The first one to be built was the **Thames and Severn Canal**, opened in 1789 from the Thames near Lechlade to Stroud, where it connected with the **Stroudwater Canal**, an eight mile link between Stroud and the river Severn opened in 1779. Despite the obvious advantages of its route, the Thames and Severn never prospered.

It suffered from shortage of water, it was poorly constructed and its locks were too small, making it impossible for the large trows, or barges, of the Severn to sail directly to the Thames. All cargoes had to be transhipped on to smaller barges at Brimscombe, where a major inland port grew up to serve the canal. The journey between the two rivers was slow and laborious, and so the Thames and Severn carried little traffic, particularly during the latter half of the nineteenth century. The last through journey was made in 1911, after which the canal gradually decayed until its final closure in 1933. The Stroudwater was closed a few years later,

The recently restored portal of the 3000 yard-long Sapperton Tunnel in Gloucestershire, the most impressive feature of the long-closed Thames and Severn Canal.

Hidden delights lie in store for those who take the trouble to explore England's abandoned canals, in this case the Thames and Severn near Cirencester.

in 1941. Although sections of the canal have disappeared, it is possible today to follow its route. Indeed, a Trust has been established to campaign for its reopening and already considerable progress has been made. An easy section to follow is between Stroud and Chalford, where there is a clear towpath and the remains of locks, mills and other canal structures. However, the most memorable features of the Thames and Severn are the 3800 yard long Sapperton Tunnel, whose ornamental portals are still to be seen, and the distinctive circular lock-keeper's houses, five of which survive. The most striking of these still stands guard over the abandoned lock chamber where the canal used to join the Thames, near Inglesham.

The Kennet and Avon Canal and its connections

The second of the great cross-country routes linking the Thames with the Severn was the **Kennet and Avon Canal**, which was completed in 1810. This was formed from three distinct waterways: the river Kennet navigation from Reading to Newbury, opened in 1723; the river Avon navigation from Bath to Hanham, near Bristol, opened four years later; and the canal built in 1810 to link the two.

The Kennet and Avon is a particularly attractive canal, engineered on a grand scale with wide locks and elegant aqueducts. It passes through a landscape of continual variation, its route linking many towns of great architectural interest, for example Bath, Bradford-on-Avon, Devizes, Hungerford and Newbury. Although successful in its early years, the canal later suffered from railway competition and

Restoration work on the Kennet and Avon Canal, where a massive flight of 29 locks at Devizes is slowly coming to life thanks to dedicated volunteers.

trade declined steadily through the latter part of the nineteenth century. Its condition also deteriorated, and the last through passage was made in 1951. Although never formally closed, the canal rapidly became derelict, until the Kennet and Avon Canal Trust was formed to campaign for its restoration. During the last 20 years remarkable progress has been made and much of the route is now open again, thanks largely to the volunteer efforts of the Trust and its members. Today only a few major obstacles remain, the mighty flight of 29 locks at Caen Hill, near Devizes, and a number of places where main roads cross the canal on the level. However, it is now clear that by the mid 1980s boats will once again be able to travel from the Thames to Bristol overland. Although parts of its route are remote, the Kennet and Avon is easy to explore as it has a good towing path throughout its length. It also has a number of remarkable features, including the Caen Hill flight, the aqueduct at Limpley Stoke, the steam pumping engine at Crofton and the tunnel beneath Savernake Forest, as well as a wealth of interesting buildings.

In its heyday the Kennet and Avon was itself part of a larger waterway network. At its western end, near the Dundas Aqueduct at Limpley Stoke, are the remains of the junction with the short **Somerset Coal Canal**. Opened in 1805, this was built to link the Somerset coalfield with the rest of the inland waterway network. It was a dramatic undertaking, with flights of locks, inclined planes, tunnels, aqueducts and other engineering features and during the 1830s and 1840s it carried over 100,000 tons of coal a year, from over 20 pits. It remained in use until the 1890s when much of its course was built over by railway companies. Today, its route can still be traced, and some of its main features, the Midford aqueduct and the locks at Combe Hay for example, can still be seen and are reminders not only of the canal, but also of a period when coal mining was a staple industry in Somerset.

The third cross-country waterway was the **Wiltshire and Berkshire Canal**, a long-forgotten canal that meandered from Abingdon on the Thames to its junction with the Kennet and Avon at Semington, west of Devizes. The Wiltshire and Berkshire Canal was opened in 1810, a remote, rural undertaking with no major engineering features, designed to supply an alternative route for the Somerset coal traffic to Oxford and the Midlands. Its route managed to avoid all major towns, although branches were built to Calne, Chippen-

*Brunel's dramatic suspension bridge, which crosses
the Avon gorge at Clifton, near Bristol,
high above the river.*

ham and Wantage. Surprisingly it remained quietly profitable until the 1870s but then it declined rapidly under the pressure of railway competition and was finally abandoned in 1914. Because its route was largely through undeveloped farmland, much of the canal can still be traced. A good place to start is Wootton Bassett where there are several old locks close together, while its course is quite distinct near Wantage. In 1819 the canal was linked to the Thames and Severn by a short branch, the **North Wiltshire Canal**, which helped to increase its traffic. Although long abandoned, this can also be traced, particularly around Cricklade, where there are the remains of two aqueducts, one over the Thames, and the old canal basin at Latton, where the canal joined the Thames and Severn. In all, there are over 65 miles of forgotten canals in Wiltshire and Berkshire, an interesting legacy of the 'canal mania' of the early nineteenth century.

The Bristol Avon
The **Avon** rises south of Malmesbury and flows southwards through Chippenham and Melksham in a typical Wiltshire landscape. It is an area rich in fine houses, including Lacock Abbey with its museum of photography, the moated Great Chalfield Manor

and Westwood Manor with its topiary garden. At Bradford-on-Avon the river swings westwards and is joined by the Kennet and Avon Canal which shares its valley course as far as Bath. From Bath to Bristol the Avon is a busy river navigation, passing through Bristol's floating harbour. Formerly one of Britain's major ports, Bristol sees little commercial traffic now, but Brunel's *Great Britain*, the first screw-driven transatlantic steamship, is preserved in the harbour as a fitting reminder of the city's maritime past. West of Bristol the river, now tidal, sweeps through the Clifton Gorge, far below Brunel's dramatic suspension bridge and then flows quickly through an industrial environment to Avonmouth docks, and the wide expanse of the Severn estuary. The river Avon represented the final link in the chain of navigations that enabled boats to travel from London across England to Bristol.

The Wey and its connections
The **Wey** has two sources, both of which drain the South Downs. One rises near Alton in Hampshire and flows through Farnham and past the ruins of Waverley Abbey to Milford, where it joins the other arm. This rises in Sussex, south of Frensham Ponds, some of the many ponds in this area formed originally to supply water power for the iron smelting industry of the sixteenth and seventeenth centuries. The Wey flows through typical Surrey countryside to Godalming, where it becomes navigable. Built originally by the Weston family of

*A view of the Wey at Weybridge, the start of this
attractive river navigation's rural route to
Godalming in Surrey.*

Winter on the Grand Union Canal in Buckinghamshire, a reminder of the days when commercial carrying by narrow boat was an all-year-round activity.

Stoke Bruerne in Northamptonshire. A traditional canal village, it is the home of the Waterways Museum, one of the most important canal museums in England.

Sutton Place, the Wey navigation was opened from Guildford to the junction with the Thames at Weybridge in 1653. Eighty years later it was extended to Godalming. The Wey is a pleasant, rural navigation, relatively unchanged since the nineteenth century with riverside mills, quiet villages and it retains an air of privacy even when surrounded by London's suburbia. Guildford is the only town of any size, and it makes the most of its river. Less appealing towns, Woking, Byfleet and Weybridge, for example, are conveniently avoided by the river. There are a number of attractions on or near the Wey: Tudor Sutton Place; eighteenth-century Clandon House; the Royal Horticultural Society gardens at Wisley; and the remains of the old Brooklands motor racing circuit. Its greatest appeal, however, is its unchanged quality. The Wey is still navigable, although commercial traffic ceased in the late 1960s. Now owned and operated by the National Trust, the Wey is a popular cruising river.

In the nineteenth century the Wey was even more important. It was the first part of a long and tortuous navigation that linked London with Portsmouth and the English Channel (for details see *The South*). It also connected Basingstoke with the Thames via the **Basingstoke Canal**. This derelict waterway joins the Wey at West Byfleet, having followed a rural and heavily locked route. Easily explored by car or on foot, the Basingstoke Canal is slowly being restored and will eventually be open to boats as far as Odiham. The original terminus at Basingstoke is permanently separated from the rest of the canal by the collapsed Greywell tunnel.

London's canals

Apart from the Thames, there are, or were a number of other waterways in London. Rivers such as the **Fleet** and the **Tyburn** have long since disappeared or been converted into underground sewers. Some canals have also vanished but not without trace. Much of the route of the optimistically named **Grand Surrey Canal** which ran from the Thames at Rotherhithe to Camberwell and Peckham can still be traced, for this was part of the Surrey Docks complex and was not finally closed until 1970. More obscure is the **Croydon Canal** which opened in 1803 to link Croydon with the Grand Surrey Canal at New Cross. The canal had 26 locks in 9 miles and it was closed in 1836 after an uneventful life. However, a length of this unlikely waterway can still be seen in Anerley, converted to a park.

London still has a number of active canals, mostly linking the Thames and the London docks with the **Grand Union Canal**. Conceived as a more direct route between London and the Midlands, the Grand Union, or rather Grand Junction Canal as it was originally called, was opened at the start of the nineteenth century. Its route was from Braunston, where it joined the Oxford Canal to Brentford in west London, where it joined the Thames. With its large-scale engineering and its more direct route linking major towns, the Grand Junction Canal was immediately successful, establishing a pattern of busy commercial carrying that continued until the 1960s. Today, the Grand Union is still the major waterway link between London and the Midlands and the North, but pleasure boats now make up its traffic. However, plenty of the old narrow boats survive and these can often be seen on the Grand Union, sometimes still in their old working formation of paired boats.

The Grand Union's route north from Brentford is predominantly suburban and industrial but once north of Uxbridge, having passed the branch to Slough, the environment improves. There are plenty of locks, some grouped in flights, often with convenient canalside pubs, as the canal climbs

towards its summit level at Tring. The canal is particularly attractive in Watford, King's Langley and Berkhamstead, while Marsworth offers an old canal workshop, branches to Wendover and Aylesbury, and nearby a variety of unusual birds on Tring Reservoirs. Between Tring and Bletchley the canal passes through rolling, open countryside, flanked by attractive villages, and a number of grand houses, including Mentmore, Ascott and Woburn. North of Bletchley the canal is enjoying a new lease of life as part of Milton Keynes. At Cosgrove is the junction with the old branch to Buckingham. This was closed in 1910, but it is still possible to follow much of the

This aqueduct carrying the Paddington arm of the Grand Union Canal across the North Circular Road provides a dramatic lesson in transport history.

route on foot. A few miles further north is Stoke Bruerne, a traditional canal village and the home of the Waterways Museum which brings to life over 200 years of canal history. Nearby is Blisworth tunnel, over 3000 yards long and opened in 1805. It is just wide enough for two boats to pass, but there is no towing path in the tunnel. Just beyond the tunnel is Gayton, and the branch to Northampton, which is a vital link between the canal network and the river Nene, and thus with the rivers and waterways of eastern England. The canal then travels through a

The wide locks of the Grand Union Canal at Berkhampstead, which carry the waterway

up to its summit at Tring. The landscape is a blend of countryside and suburbia.

more remote landscape, crossing the town of Weedon on a huge embankment, and passing the junction with the Leicester canal, to the east Midlands and the Trent, before entering the 2000 yard Braunston Tunnel. Braunston itself is an unchanged canal village and a good place to see traditional canal boats. Shortly after the village the Grand Union meets the Oxford Canal, and then follows a varied route to Birmingham. It passes Warwick and Leamington; the dramatic flight of 21 locks at Hatton; Knowle; and the Birmingham suburbs, mostly hidden by the wooded sides of a deep cutting. This completes the direct route from London to the Midlands.

In order to expand its trade in London, and to avoid the problems caused by the tidal Thames, the Grand Junction company built a branch canal to Paddington which opened in 1801. This swings round western London in a broad curve to meet the **Regent's Canal** at Little Venice, a picturesque canal centre in the heart of London. The Regent's Canal was opened in 1820 to connect the Paddington Branch of the Grand Union with London's dockland and the Thames via the Limehouse Basin. This canal has many elegant and interesting features and is one of the least known pleasures of London. It passes through Regent's Park and the Zoo; Camden Town; round the back of St Pancras; under Islington in a long tunnel; through Hackney and on to London's East End. Much of the towing path has now been opened as an urban footpath and the best parts of the canal are easily accessible. In addition, there are a number of regular waterbus services. An interesting way to see London is by water, via the Thames, the Grand Union Canal, the Paddington Branch and the Regent's Canal.

The Lee and the Stort

The last of London's river navigations is the **Lee**. This rises near Luton and flows through St Albans, and Welwyn to Hertford, where it becomes navigable. The Lee is a river of great contrasts. Its upper reaches through rural Hertfordshire are attractive, linking interesting towns and villages, such as Ware, Stanstead Abbots, Broxbourne and Waltham Abbey. It then becomes a pleasantly suburban river, before plunging through the industrial and urban sprawl of north-east London. Much of the Lee valley is now being developed as a centre for sport and recreation and it is a rural lifeline into the centre of London. The Lee joins the Thames at Limehouse Basin, where it also connects with the Regent's Canal and the rest of the canal network. There is also a short branch, the Hartford Union Canal, which links the Lee directly with the Regent's Canal. The lower reaches of the Lee still see some commercial traffic, although the main users of the navigation are now pleasure boats.

The river **Stort** was made navigable in the 1760s, to link Bishop's Stortford with the river Lee, and thus with the Thames. Dependent for many years on agricultural traffic, the Stort is still one of the least developed and most attractive of the waterways of southern England. Its landscape has East Anglian qualities, and its route is marked by attractive villages and waterside mills, notably at Sawbridgeworth. The Stort bypasses much of Harlow New

Dramatic urban scenery on the Regent's Canal near Paddington in west London, where a short arm of the canal leads into Paddington Basin.

The elegant mill and maltings at Stratford on the river Lee, one of a number of stylish industrial buildings along the busy lower reaches of the Lee.

Town but its quiet and privacy may in future be affected by the expansion of Stanstead into London's third airport.

The Darenth and the Roding

Two of the least known and most attractive of the Thames tributaries are the **Darenth** and the **Roding**. Both are surprisingly remote and rural for much of their course, despite their proximity to London. The Darenth rises near Westerham in Kent, and flows along a pleasant wooded valley to Sevenoaks. It then turns north, passing Otford, Eynsford and Farningham on its way through the Downs. Despite its rural nature, it is a river of some history. There is a Roman villa at Lullingstone,

castles at Eynsford and Lullingstone, and the remains of a Bishop's Palace at Otford. North of Swanley the Darenth loses its rural quality, and flows through Dartford and the Erith marshes before joining the Thames. The Roding is in many ways the Essex equivalent of the Darenth. It is a quiet river, meandering for much of its course through attractive farmland and remote villages. Famous once for its eels, the river now adds a traditional flavour to London's outer suburbs. Mills, fine churches and farms accompany the Roding as far as Loughton, where it surrenders its privacy to the MII motorway and suburban development. At Ilford the Roding becomes tidal and then joins the Thames at Barking Creek.

For much of its course the river Wye flows quickly through a steep and wooded valley that is undoubtedly one of the finest in England. It is seen here near Symonds Yat.

THE SEVERN

The Severn is one of England's greatest rivers and it has determined for centuries the social and economic structure of its region. Effectively, its wide and fertile valley forms the border between England and Wales, providing a more powerful and permanent barrier than the castles and fortifications erected by man. Between the Roman period and the late eighteenth century the Severn was navigable far north of Shrewsbury and deep into Wales, and so it was the backbone for all development of the area. Its importance can be measured in many ways. Great cathedrals, abbeys and castles line its banks and a series of inland ports, Gloucester, Worcester, Stourport, Bewdley and Ironbridge, mark its route. Rich agricultural lands surround it and the industries associated with the river since the Middle Ages; coal, iron, milling, pottery and porcelain, boatbuilding and fishing; are all part of the Severn's history.

The importance of the river is also evident from its huge network of connecting waterways, many of which serve the inland ports along the Severn. Rivers which join the Severn include the Wye (to Hereford), the Teme (to Ludlow) and the Avon (to Stratford), while canals, built mostly during the late eighteenth century, connected many parts of the Severn with the English waterways network. Many of these canals can still be explored today and some of the inland ports, notably Sharpness and Gloucester, are still in commercial use. As a result, boats can use the Severn to visit many parts of western England and the Midlands.

The scenery of the Severn valley is very varied. Rolling farmland, orchards, steep wooded valleys

The imposing ruins of Buildwas Abbey in Shropshire. The church, built in the twelfth century, is the best preserved part of the abbey.

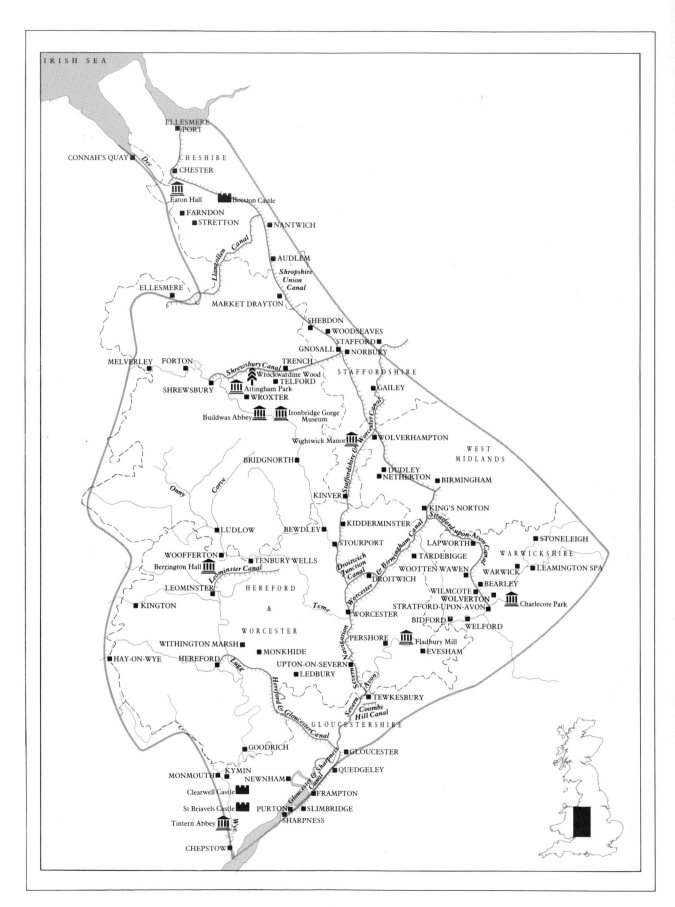

IRISH SEA

ELLESMERE PORT

CONNAH'S QUAY

CHESHIRE

CHESTER

Dee

Eaton Hall

Beeston Castle

FARNDON

STRETTON

NANTWICH

Llangollen Canal

AUDLEM

Shropshire Union Canal

ELLESMERE

MARKET DRAYTON

SHEBDON

WOODSEAVES

STAFFORD

GNOSALL

NORBURY

MELVERLEY

FORTON

TRENCH

STAFFORDSHIRE

Shrewsbury Canal

Wrockwardine Wood

TELFORD

GAILEY

Attingham Park

SHREWSBURY

WROXTER

Buildwas Abbey

Ironbridge Gorge Museum

Staffordshire & Worcester Canal

Wightwick Manor

WOLVERHAMPTON

WEST MIDLANDS

BRIDGNORTH

DUDLEY

NETHERTON

BIRMINGHAM

Onny

Corve

KINVER

KING'S NORTON

LUDLOW

BEWDLEY

KIDDERMINSTER

Stratford-upon-Avon Canal

STONELEIGH

WOOFFERTON

STOURPORT

LAPWORTH

WARWICKSHIRE

TENBURY WELLS

Droitwich Junction Canal

TARDEBIGGE

LEAMINGTON SPA

Berrington Hall

Leominster Canal

WOOTTEN WAWEN

WARWICK

BEARLEY

LEOMINSTER

HEREFORD

Worcester & Birmingham Canal

DROITWICH

WILMCOTE

KINGTON

&

Teme

WORCESTER

WOLVERTON

Charlecote Park

WITHINGTON MARSH

WORCESTER

STRATFORD-UPON-AVON

MONKHIDE

BIDFORD

WELFORD

HAY-ON-WYE

HEREFORD

PERSHORE

Fladbury Mill

Lugg

Severn Navigation

EVESHAM

UPTON-ON-SEVERN

LEDBURY

Hereford & Gloucester Canal

Severn

Avon

TEWKESBURY

Coombe Hill Canal

GLOUCESTERSHIRE

GOODRICH

GLOUCESTER

KYMIN

QUEDGELEY

MONMOUTH

NEWNHAM

Gloucester & Sharpness Canal

FRAMPTON

Clearwell Castle

St Briavels Castle

PURTON

SLIMBRIDGE

Tintern Abbey

Wye

SHARPNESS

CHEPSTOW

The suspension bridge across the mighty estuary of the river Severn, a seemingly delicate but vital link between England and Wales.

and dramatic hills can all be found in close proximity. The towns and villages of the Severn valley exhibit interesting variations in regional architecture, from the timber-framed styles of Shropshire and Cheshire to the Georgian elegance of Shrewsbury and Upton. There are also fine churches, interesting country houses, and a rich legacy of industrial history that reflects the part played by the Severn in the development of the Industrial Revolution. The river's industrial history can be explored in areas such as the Forest of Dean, and particularly at Ironbridge, the cradle of modern industrialisation. The passage of time has softened these relics of industry and given them picturesque qualities. The region is now predominantly rural, although modern industry is never far away, with the Black Country and the Midlands spreading to the east. Here again history can be explored in an unusual way, however, for the Black Country still boasts over 100 miles of navigable canals, the remains of a huge network that was the lifeblood of the region during the nineteenth century. There are several connections between the Severn, and the canals of Birmingham and the Midlands. These little known waterways mark the contrast between the rural and the industrial landscapes and bring to life a vibrant era of English history.

To the north of the Severn lies Chester and the river Dee. Chester is one of England's finest small cities, endowed with a rich collection of typical Cheshire buildings. This area enjoys a varied landscape of farmland interspersed with the ponds and low-lying marshes that are the result of centuries of salt extraction. A number of traditional market towns are linked to the Severn by the Shropshire Union Canal. This, one of the last main line canals to be built in England, follows an attractive route from Wolverhampton to Ellesmere Port.

The Severn

The **Severn** rises in Wales, high in the Cambrian Mountains south of Machynlleth, and sweeps in a series of dramatic bends through Llandidloes and Newtown to Welshpool. From here it swings east towards Shrewsbury, crossing the border into England by the village of Melverley. Although hard to believe today, Welshpool was the limit of navigation early in the nineteenth century, when the typical

Abraham Darby's iron bridge over the Severn. Built in 1778, it was the first iron bridge in the world and underlines the importance of Coalbrookdale.

Severn sailing barges, or trows, regularly made the 128 mile journey upstream from Gloucester. During the last 150 years the navigable section of the river has steadily dwindled, and now only about 45 miles from Gloucester to Bewdley is accessible to other than small, portable boats.

The Severn is a wide and generous river contained by high banks, which winds its way through attractive, and sometimes dramatic scenery. Its long history as a navigation stretches back to the Roman period and probably beyond, and this ensured that a number of fine towns have grown up along its banks: Shrewsbury, Bridgnorth, Bewdley, Worcester, Upton-on-Severn, Tewkesbury and Gloucester. Not many rivers can boast two cathedrals and a major Norman abbey within a few miles. The Severn also connects with a number of other waterways, making it an important through route for cruising, and in its lower reaches, for commercial traffic. The character of the Severn is quite distinct. The first impression is one of size. Even the now unnavigable section between Bewdley and Welshpool seems wide and impressive, far more important-looking than the equivalent upper reaches of the Thames which, by comparison dwindles to a stream quite quickly above Lechlade.

It also seems relatively untamed. The high banks serve as permanent reminders of the power of the river when in flood, and the six locks on the navigable section are large and powerful structures, clearly built to handle small ships rather than boats or barges.

The route taken by the Severn is well worth exploring, for there is much to be discovered along its banks. Between the Welsh border and Shrewsbury the river is rather remote. Few roads approach its banks but the landscape is interesting, with distant views of the Long Mynd to the south. The Severn really comes to life when it reaches Shrewsbury, a town that makes the most of its river. Here, there are fine bridges, a castle, several good churches, timber-framed and Georgian buildings and museums, while to the east of the town are the twin attractions of eighteenth-century Attingham Park and Roman Wroxeter. Approaching Ironbridge, the Severn enters a steep wooded gorge, overlooked to the north by the Wrekin. It is hard today to realise that this area was one of the starting points for the Industrial Revolution. Here, at Coalbrookdale, Abraham Darby first smelted iron with coke, the revolutionary technique that was to make cast iron the raw material for industrialisation. In the eighteenth century the valley was filled with the smoke and flames of the furnaces, a scene depicted by many painters of the period as a vision of Hell. Today Coalbrookdale is a wooded valley filled with picturesque and overgrown ruins, memorials to England's industrial past, now all well cared for by the Ironbridge Gorge Museum Trust. There are furnaces and factories, railways and canals, porcelain and tile works, wharves and quays. In the centre is the great iron bridge itself, the first iron bridge in the world, crossing the river in the splendid steep arch that has stood since 1778. There are also many eighteenth- and nineteenth-century buildings of interest, while the ruins of Buildwas Abbey connect the area with the twelfth and thirteenth centuries.

The wooded valley of the Severn continues to Bridgnorth, a handsome town with a castle and, somewhat unexpectedly, a funicular railway. Bridgnorth has other railway connections, for it is the northern terminus of the Severn Valley Railway, which is one of the premier steam railways in England and a fine way to see the river between Bridgnorth and Bewdley. Although the trows and barges have long departed, Bewdley still has the

An enjoyable way to explore the river Severn is by steam train. The route of the Severn Valley Railway includes this iron viaduct over the river at Arley.

atmosphere of an eighteeenth-century inland port, with fine quays and warehouses, and a maze of narrow streets to be explored on foot. A similar but rather more down-to-earth atmosphere pervades Stourport, a town developed entirely in the late eighteenth century around the junction of the Severn and the Staffordshire and Worcestershire Canal. The large canal basins with their elegant buildings and bridges are an indication of the scale of traffic that formerly used this waterway route from Stafford via Wolverhampton and the Black Country to the Severn. Although still in a valley, the river is flanked by a more open landscape between Stourport and Worcester. Worcester is also an excellent river town, with a riverside dominated by the cathedral and the eighteenth-century stone bridge. Here, large locks lead from the Severn into the extensive network of basins that serve the Worcester and Birmingham Canal, another eighteenth-century link between the Severn and the Black Country.

An open landscape of meadows and woods follows the river south towards Tewkesbury and Gloucester, with distant views of the Malvern Hills to the west, and Bredon Hill and the Cotswolds to the east. Along the way is Upton, another fine Georgian town, its riverside dominated by the church with its curious copper dome. At Tewkesbury is the Norman abbey and the confluence of the Severn and the river Avon. This part of England is famous for its fruit, and the Severn is very much a cider river here, with a number of pubs along its banks specialising in the many different varieties and flavours. From Wainlode Hill, between Tewkesbury and Gloucester, there is a good view of the Severn, now a wide and mighty river winding its way through a landscape of rolling farmland. Nearby is the entrance to the disused **Coombe Hill Canal**, a relic of an eighteenth-century plan to build a canal from the Severn to Cheltenham. Only three miles, with two locks, were ever built and it was closed in 1876.

North of Gloucester the Severn divides into two channels. One leads to the old **Severn Navigation** that followed the original course of the river to Sharpness and the estuary. The other passes through Gloucester and its extensive docks and basins to meet the **Gloucester and Sharpness Canal**, a wide ship canal opened in 1826 to bypass the difficult stretch of the Severn. The old route, which is tidal throughout, follows a wild course through marshland, the river flowing steadily faster among sandbanks to its estuary. This is a remote area, noted for its wildlife. At Slimbridge the Wildfowl Trust has one of the largest reserves in the country. It is also a good place to see the Severn

Worcester Cathedral, with its magnificent fourteenth-century tower, dominates the river Severn. The city was granted its first royal charter in 1189.

Bore, a tidal wave that sweeps up the Severn from the Atlantic about 250 times a year. It is at its best during the high tides of spring and autumn, when waves up to nine feet high drive up the river, briefly reversing its natural flow. One of the best viewpoints is at Stonebench, west of Quedgeley. South of Sharpness the estuary rapidly widens as the Severn becomes a major seaway, passing Chepstow, the mouth of the Wye, the Severn road bridge and the huge docks complex at Avonmouth, the modern port of Brisol.

The 16 mile long ship canal follows a more sedate and rural route and is still used by a variety of ships and barges making their way from the Severn estuary to Gloucester's docks, and occasionally northwards to Worcester. All the bridges over the canal can be opened, allowing the passage of ships

up to 190 feet long and 29 feet wide. Timber and oil are the principle cargoes. The canal itself is not unattractive, but there are few towns or villages along its route. The most interesting features are at Sharpness, where there are the extensive and busy docks, at Purton, where the canal overlooks the Severn, and at Frampton, which claims to have the largest village green in England. Near Gloucester is Saul Junction, a former waterway crossroads, where the **Stroudwater Canal**, closed since 1954, crossed the Gloucester and Sharpness on its way to join the Thames and Severn Canal. Until the decline of the latter in the late 1920s, it was possible to travel from Lechlade on the Thames direct to the Severn via the Stroudwater Canal, or to Gloucester and the other Severn towns. Gloucester's docks are well worth exploring, as many of the original nineteenth-century wharves and warehouses still survive. Fine and elegant red-brick buildings, many of them are now either disused or converted to other purposes.

The Severn's surviving connections

In its heyday, the Severn was at the heart of a network of connecting waterways that linked the river with the Midlands, London, the South-West, South Wales, North Wales, Liverpool, Manchester, and ultimately with all the other inland waterways of England. A great many of these waterways have disappeared, but a number still survive that can easily be followed and the whole of the Severn region is well worth exploring. The connections that still survive are the **Worcester and Birmingham** and **Staffordshire and Worcestershire** Canals, and the river **Avon Navigation**. The Worcester and Birmingham Canal was opened in 1815, linking Gas Street Basin in the heart of Birmingham with the Severn at Worcester. The canal runs for most of its route through attractive countryside, but it is most notable for its engineering. Its 30 mile length includes 5 tunnels and 58 locks, of which 36 are grouped together in less than 3 miles, between Tardebigge and Stoke Prior. Nowhere in England is there a greater flight of locks. In addition, the flight starts at Tardebigge with the deepest lock (it has a 14 foot drop) in the English narrow canal network. Of the tunnels, the longest is King's Norton, just outside Birmingham, 2,726 yards of subterranean passage carved through Wast Hill. At King's Norton there is also a junction with the Stratford Canal.

Little used today, Gloucester docks are a splendid memorial to the Victorian period, and to the great days of the Severn as a commercial waterway.

The Staffordshire and Worcester Canal is a far older undertaking, and dates back to the 1770s, when it was planned by the engineer James Brindley as part of the original Grand Cross canal system to link the Thames, the Severn, the Trent and the Mersey. Despite its age, it remained in commercial use until the late 1950s. The canal runs for 43 miles from Stourport on the Severn to Great Haywood near Stafford, where it joins the Trent and Mersey Canal. It is a route that is still unusually attractive despite the proximity of the Midlands and the Black Country. A particularly old-fashioned canal, still largely unchanged since the eighteenth century, it has attractive bridges and locks, with tunnels and cuttings carved through red sandstone. The best features are the basins at Stourport, the route through the centre of Kidderminster and the remains of distinctive circular toll houses at Stewponey and Gailey. Also worth seeing are the flight of locks with their octagonal toll house at The Bratch, the village of Kinver and the dramatic Kinver Edge, Wightwick Manor with its William Morris associations and the Elizabethan gatehouse at Tixall,

overlooking a stretch of canal turned into an ornamental lake. At Stourton there is a junction with the Stourbridge Canal, which leads to Birmingham via the massive Dudley and Netherton tunnels, while a former branch, closed since the 1920s, linked the canal to Stafford.

The river Avon (not to be confused with the river that flows from Bath and Bristol to the Severn estuary at Avonmouth) links Tewkesbury on the Severn with Stratford-on-Avon. This 44-mile navigation represents one of the most remarkable achievements in recent waterway history. Rising near Stanford, south of Lutterworth, the Avon follows a meandering course through the Midlands. Passing north of Rugby and south of Coventry it avoids the main areas of industry, and manages to remain largely rural. Apart from Stoneleigh, the site of the annual Royal Agricultural show, the first place of interest is Leamington Spa. Nearby is Warwick, whose castle enjoys a splendid riverside site. The Avon is quite a substantial river, flowing through the town and under an elegant aqueduct which carries the Grand Union Canal. Between

Warwick and Stratford the Avon winds through farmland, passing the Elizabethan Charlecote Park. At Stratford, whose waterfront is dominated by the Shakespeare Memorial Theatre and Holy Trinity Church and where the river flows under two multi-arched stone bridges, the Avon Navigation starts.

The Avon was first made navigable during the late 1630s. In 1717, the river was separated into two navigations, the Upper and the Lower, which met at Evesham. Commercial traffic continued until 1873 when the river became impassable to boats larger than skiffs or canoes. The locks and weirs collapsed on the Upper river, but the Lower river remained in occasional use until the 1940s. In 1949 a Trust was formed to restore the Lower river to navigation, a task that was completed in 1965 when the river was reopened from Tewkesbury to Evesham. Work was carried out largely by voluntary labour and all the necessary finance was raised privately by the Trust. The success of this restoration scheme inspired, first a similar restoration of the Stratford Canal, from Lapworth to Stratford, and second the reopening of the Upper Avon. This was an astonishingly ambitious scheme, requiring the creation of nine new locks, numerous weirs and other works, the raising of several hundred thousand pounds and the extensive use of volunteer labour over a long period. Construction work started in 1969 and in 1974 the Avon Navigation was reopened, linking Stratford with the Severn for the first time for over a century, and making available to pleasure boats a cruising circle comprising the Avon, the Stratford Canal, the Worcester and Birmingham Canal and the Severn. The route of the Avon to Tewkesbury is attractive and rural. It winds its way through fields and between tree-lined banks from the crowded waterfront at Stratford, through Welford, Bidford with its fifteenth-century bridge, Evesham, Fladbury Mill, Pershore with its abbey and Georgian houses, and then round Bredon Hill, through Tewkesbury and on to the Severn, where both commercial barges and seagoing yachts can be seen.

The Great Parlour, Wightwick Manor. The house, extravagantly decorated by William Morris, stands beside the Staffordshire & Worcestershire Canal.

Fladbury Mill on the river Avon, one of the features that makes this recently reopened river navigation so appealing to visitors.

The Severn's lost connections

Its river and canal connections make the Severn one of the liveliest waterways in England, but a century ago it was far busier for a number of important connecting waterways have largely disappeared. These include the **Shrewsbury Canal** and the **Shropshire Canals**, the **Droitwich Canals**, the **Hereford and Gloucester Canal** and, although its connection with the Severn was never more than a dream, the **Leominster Canal**. Among the large-scale exhibits preserved by the Ironbridge Gorge Museum Trust are some short stretches of canal and an inclined plane, the remains of an extensive network of small tub-boat canals that grew up to serve the industries of Ironbridge and its region during the eighteenth century. Between 1768 and 1792 several short canals were built independently for the coal and iron traffic and in 1796 the Shrewsbury Canal was opened to link them together and to provide a connection with the Severn. These canals were small, but they required dramatic engineering. There were many locks, tunnels, and a number of inclined planes to connect the different levels and during the early nineteenth century the whole area must have been a hive of extraordinary activity. Parts of the system were closed in the nineteenth century, but much survived into this century, the network dying gradually from the 1900s as coal mines were exhausted or the ironworks abandoned. The last section to remain in commercial use was part of the Shrewsbury Canal, which was not finally closed until 1944, but the network

The inclined plane was much favoured by eighteenth-century engineers. None are in operation today but the Hay Incline, seen here, has been restored at the Ironbridge Gorge Museum.

had already been isolated by the closing in 1921 of the Trench inclined plane, the last to operate in England. The rapid expansion of this region since the formation of Telford new town has obliterated most of these canals, but traces still remain. The angled slopes of inclined planes can be found at Wrockwardine Wood and Trench, there is an aqueduct at Dawley and warehouses and other canal buildings at Wappenshall. However, the best way to understand this forgotten network is to visit Ironbridge Gorge Museum. Here are preserved boats, sections of canal, a cast iron aqueduct that formerly carried the Shrewsbury Canal over the river Tern, and an inclined plane complete with track to show how the small tub boats or their cargoes were hauled from one level to another.

In 1771 a short canal, the **Droitwich Barge Canal**, was opened to link Droitwich with the river Severn. Nearly seven miles long and with a number of locks, the canal was built to serve the salt trade, which continued to flourish throughout the nineteenth century. In 1853 another short canal, the **Droitwich Junction Canal**, was opened to join Droitwich to the Worcester and Birmingham Canal at Hanbury Wharf, thus completing a small circuit of waterways. Little used during this century, the canals were abandoned in 1939. Although built over or filled up in places, the routes can still be followed, a process that will become steadily easier as the Trust formed in 1973 to reopen the canals works towards its goal of full restoration.

One of the more unlikely products of the 'canal mania' of the late eighteenth century was the **Hereford and Gloucester Canal**, a waterway that has been largely forgotten and whose remaining artefacts stand in splendid isolation in magnificent countryside, looking as remote today as the creations of some prehistoric civilisation. The canal was built in two sections; the first, from Gloucester to Ledbury opened in 1798; the second, from Ledbury to Hereford in 1845. By the time it was completed, the canal was already out of date and what little traffic it carried was already under threat from the railway. Despite this, it somehow managed to survive and boats continued to carry building materials, coal, grain and cider, from Hereford to and from the Severn. The last commercial cargo was carried in 1883. A railway was then built on the route of the canal from Ledbury to Gloucester and this in turn has vanished, leaving little trace of Hereford's association with the industrial revolution. However, many of the buildings and engineering features that were part of the canal between Hereford and Ledbury can still be found, silent witnesses to the misplaced enthusiasms of eighteenth-century speculators. There are wharf buildings at Withington Marsh, Kymin and elsewhere, there is a pretty canal bridge at Monkhide, and Ashperton Tunnel still stands in its deep cutting. Near Prior's Court there is an aqueduct over the river Leadon. However the most significant memorial to the canal is at Oxenhall, where the great tunnel, over 2000 yards long, still survives.

Even more inconceivable now is the **Leominster Canal**, a bizarre late eighteenth-century scheme to link rural Herefordshire with the Severn. The canal was planned to run from Kington to Stourport via Leominster, a route that involved 4 major tunnels, 2 large aqueducts, and over 60 locks. Only about 18 miles, from Leominster to Southnet Wharf, were ever built before the money ran out, leaving the canal isolated and virtually useless. Some form of trading on the canal continued until it was closed in 1859, but it was virtually dead from the moment it opened. A number of relics remain to interest the explorer, for example Putnal Field Tunnel, near Berrington Hall, locks and wharves at Woofferton and, nearby, an aqueduct over the river Teme. Near Newnham there is a small tunnel, a handsome brick aqueduct over the river Rea and fine buildings at Marlbrook Wharf. However, the most intriguing memorial to the canal is the 1,250 yards long Southnet Tunnel, which was never used commercially. It was completed in 1795 as part of the route from Southnet onwards towards Stourport, but collapsed shortly afterwards and was then abandoned.

The Stratford-upon-Avon Canal

Indirectly part of the river Severn network, the **Stratford Canal** was opened in 1816 to link the river Avon at Stratford with King's Norton, south of Birmingham, where it joined the Worcester and Birmingham Canal. A short branch at Lapworth connects with the Grand Union Canal. The canal was successful at first and saw considerable traffic between the Severn and Birmingham via the river Avon, but the closure of the Avon Navigation in 1873 brought about a rapid decline, particularly of the lower section from Stratford to Lapworth. Impassable since the 1930s, this lower section remained somehow in existence until 1958 when closure seemed imminent. At the last moment a restoration scheme was mounted, using volunteer labour, and in 1964 the canal was reopened to Stratford, the restoration having cost exactly half the estimated cost of closure. Since then, the southern section has been administered by the National Trust.

The Stratford is a particularly attractive rural canal, much of its course is through quiet farmland. There are distinctive bridges, some in cast iron and

Unusual features of the Stratford-on-Avon Canal are the barrel-roofed lock houses, and the iron footbridges that divide to allow the passage of the towing rope.

split in the centre to allow the passage of the towing rope, pretty lock cottages and, at Wootten Wawen and Bearley, large iron aqueducts carrying the canal over roads and a railway. The locks are grouped in flights at Lapworth and Wilmcote. After its rather private rural route, the canal comes to a splendid

The imposing ruins of Monmouth Castle on the river Wye. A fast-flowing border river, its historical importance is underlined by the number of castles built along its banks.

terminus in a basin surrounded by flowers beside the Shakespeare Memorial Theatre in the heart of Stratford, a terminus that is now connected once again with the restored Avon Navigation.

The rivers of the Border

One of the most attractive of English rivers is the Wye, its longstanding popularity based on the magnificent countryside that surrounds it and the diversity of towns, villages and buildings to be found on its banks. Rising in Wales, the river crosses the border at Hay-on-Wye, a town at the foot of the Black Mountains that houses the world's largest bookshop. Between Hay and Hereford small villages accompany the remote course of the river, but after Hereford its route becomes more dramatic, winding its way through a wooded valley. A number of pretty suspension bridges cross the river, adding to the almost Alpine flavour. There are plenty of castles to underline the historical importance of the Wye as a border river: Goodrich, Monmouth, Clearwell and St Briavels. The steep wooded banks contain memorials to many other centuries and there are prehistoric monuments, associations with King Arthur, sections of Offa's Dyke and a number of traces of the industrial history of the region, notably the iron industry that flourished in the Forest of Dean and along the banks of the Wye during the seventeenth and early eighteenth centuries. The most spectacular part of the Wye is

probably the stretch from Monmouth to Chepstow and the junction with the Severn estuary, a steep and twisting valley that is a suitably romantic setting for Tintern Abbey. Today the fast-flowing waters of the Wye attract only canoeists and fishermen in small dinghies, but until the middle of the nineteenth century it was possible for barges to reach Hereford. A traditional navigation dating from the thirteenth century, the Wye was greatly improved in the 1660s and by the eighteenth century boats could find their way to Hay, hauled on long ropes by gangs of men, a technique used also on the upper reaches of the Severn. Since the navigation was abandoned in the 1860s, the river has steadily silted up, and its fast-flowing and tortuous course today shows no trace of its commercial past.

A tributary of the Wye is the **Lugg**, which flows from its source in the Radnor Forest west of Presteigne to Leominster, and then to its junction with the Wye east of Hereford. The Lugg is an undramatic rural river, its course marked by several pretty villages, but it too was once a navigation. During the eighteenth century locks were built to enable barges to reach Leominster, a traffic that continued until the 1860s, but nothing remains today.

The third border river is the **Teme**, which, with its tributaries, links another stretch of splendid countryside to the Severn. A group of rivers, the Onny and the Corve and their associated streams flow from Wenlock Edge to join the Teme near Ludlow. The Teme, which rises in Wales west of Clun Forest, becomes a significant waterway in Ludlow, a splendid town whose castle overlooks the fast-flowing river. From Ludlow the Teme continues to Tenbury Wells and thence to its junction with the Severn south of Worcester, its course marked by wooded valleys, rolling farmland, hop and fruit fields and a series of quiet and undeveloped villages. Although not a classic waterway itself, the Teme adds Ludlow to the list of the fine towns of the Borders and the Severn valley that are linked by water.

The Dee

One of the great mountain rivers of north Wales, the **Dee** rises at Lake Bala, and then follows a tortuous and often violent course among rocks and rapids to Llangollen. As it approaches England, the river becomes more sedate, the mountains giving way to rolling farmland that forms a pleasant and decorative background to its twisting course. At Farndon, east of Wrexham the Dee becomes tidal and navigable as it flows under the fourteenth-century bridge. The pleasant rural course of the river continues to Chester, passing the grounds of Eaton Hall, after which it really comes to life. Chester is a handsome city with a complete city wall, a cathedral, a wealth of timber-framed, Georgian and Victorian buildings, and a pretty waterfront with a number of attractive bridges. It boasts a variety of boats, from canoes and skiffs to large passenger launches, as well as the occasional barge, for the Shropshire Union Canal also passes through Chester and there is a lock

The castle at Ludlow is one of the many attractions of this fine town built above the flowing waters of the river Teme.

to connect the canal to the Dee. West of Chester the character of the Dee changes dramatically as it becomes a fast-flowing tidal estuary, with sandbanks and other hazards to navigation. Low-lying marshland and, near Connah's Quay, industrial development dominate the banks.

The Shropshire Union Canal

The **Shropshire Union Canal** Company was formed in 1846 from the amalgamation of a number of smaller, independent canals, dating mostly from the eighteenth century. Despite considerable reduction through closures during the 1930s and 1940s the Shropshire Union network is still extensive. The main line runs from Autherley, on the Staffordshire and Worcestershire Canal north of Wolverhampton to Ellesmere Port on the Mersey estuary, where there is a connection with the Manchester Ship Canal. There are two branches: one runs from Nantwich to the Trent and Mersey Canal at Middlewich; the other climbs through spectacular scenery to Llangollen in north Wales. The main line was built in three sections, each of which has a quite distinctive character. The first section, from Nantwich to Chester, was opened in 1774. It is a wide canal whose original eighteenth-century elegance is still apparent. The second section, opened in 1796, runs from Chester northwards to Whitby on the Mersey, later renamed Ellesmere Port. The third, opened in 1835, goes from Nantwich southwards to Autherley. This last section, handsomely engineered by Thomas Telford, reflects the improved

This elegant iron aqueduct carries the Shropshire Union Canal over the A5 near Brewood. The road and canal were both engineered by Thomas Telford.

The Boat Museum at Ellesmere Port, housed in the old basins and warehouses that formed the northern terminus of the Shropshire Union Canal.

standards of its day, with its straight route, high embankments, deep cuttings, tall bridges and locks grouped in flights.

The directness of the route between Wolverhampton and the Black Country and the Mersey ports enabled the Shropshire Union to remain in commercial use until the 1960s. Despite this, it is an attractive, predominantly rural canal, its route marked by interesting canal architecture and engineering features. Particularly notable are the bridges, the buildings and wharves, especially those at Autherley, Norbury, Tyrley, Market Drayton, Nantwich and Chester; the lock flights at Tyrley and Audlem, and the dramatic cuttings and embankments at Gnosall, Woodseaves and Shebdon (that at Shebdon took six years to build). There are also fine aqueducts that carry the canal over the A5 at Stretton and Nantwich. The main towns, Market Drayton, Nantwich and Chester, all feature the black-and-white timbered architecture of the region and are well worth exploring. The course of the canal through Chester is particularly impressive, featuring a flight of locks cut from solid rock beneath the towering city walls. Chester Zoo is by the canal north of the city. Between Nantwich and Chester the canal winds through attractive and often dramatic scenery, dominated for several miles by the imposing ruins of Beeston Castle, perched high on a rocky and overgrown hilltop. North of Chester the most interesting feature is Ellesmere Port, currently being developed into a centre for living canal history. The surviving nineteenth-century wharves and warehouses there are now the home of the Boat Museum, a large and steadily growing collection of inland waterway craft and related relics, where the skills of the canal craftsmen are being preserved for posterity against the background of large vessels passing on the Manchester Ship Canal.

At Hurleston, just north of Nantwich, the branch leading to north Wales leaves the main line of the Shropshire Union. At first a rural, meandering canal, more like a river in character, the Llangollen arm becomes more dramatic as it approaches Wales. The strange and wild landscape of the peat bogs at Whixhall Moss gives way to the so-called 'lake district' of Ellesmere, a group of nine wooded lakes that surround the canal. Ellesmere, itself, is a pleasant country town with an attractive basin lined with original warehouses, at the end of a short arm.

At Welsh Frankton is the junction with the long Montgomery branch that runs southwards to Welshpool and Newtown, a beautiful canal derelict since the 1940s, but now slowly being restored and brought back to life. A few miles further on the canal crosses into Wales over Telford's great stone aqueduct at Chirk. Nearby, at Pontcysyllte, is one of the greatest wonders of the waterways, an astonishing iron aqueduct over 1000 feet-long that strides across the river Dee on great stone pillars, 120 feet up in the air. Designed by Telford and completed in 1805, this masterpiece should be admired both from the Dee valley below, framed by Welsh mountains, and from the precipitous towpath beside the canal on the aqueduct itself.

Until its closure in 1944 there was another branch canal that connected the main line at Norbury to the Shrewsbury Canal at Wappenshall. Opened in 1835, this 10 mile waterway with its 23 locks was a vital link between the canal system of Ironbridge and the Severn valley, and the main canal network. Little of the route now remains, but part of the lock flight at Norbury, further locks at Newport and a three-arch aqueduct at Forton can still be seen.

The Birmingham Canal Navigations

In order to understand the importance of the development of the English canal system in the eighteenth century, it is vital to visit Birmingham

The Black Delph flight of eight locks, which connects the Birmingham Canal Navigations with the Stourbridge Canal, in a landscape that is full of contrasts.

An unusual view of Spaghetti Junction from below, one of many strange sights to be seen along the hidden waterways of the Birmingham Canal Navigations.

which has an extraordinary network of predominantly urban canals. The **Birmingham Canal Navigations** grew up from 1769 to serve the new industries of Birmingham and the Black Country, and at its peak in the middle of the nineteenth century the network totalled over 160 miles. There were over 200 narrow locks, more than 500 private arms or basins connecting factories to the system, 3

Farmer's Bridge locks at the start of the Birmingham and Fazeley Canal. This was once a busy life line of the Birmingham Canal Navigations.

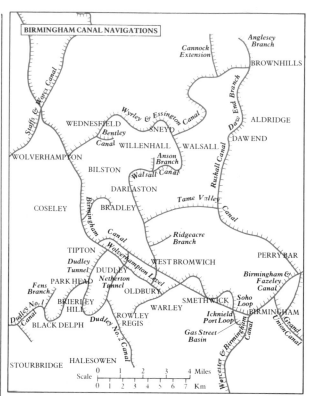

tunnels over a mile long, at Dudley, Netherton and Lappal, numerous aqueducts and bridges, and many branches connecting the network to other English canals. About 100 miles of the network still survives, allowing Birmingham to claim more miles of canal than Venice. To explore the Birmingham Canal Navigations (or BCN as it is commonly known) is to step back in time and enter a secret world, largely unchanged since the Victorian period. This exploration can only be made by boat or on foot, but the effort is worthwhile for it brings the history of the Industrial Revolution to life. Commercial use of the BCN continued until the early 1970s, but since then industry has gradually retreated, allowing much of the network to become both an unofficial museum of the Industrial Revolution and an unofficial wildlife park. It is a private world where flowers, butterflies and birds thrive among the overgrown ruins of industry and into which the twentieth century only occasionally intrudes. To travel by boat unseen beneath the concrete chaos of Spaghetti Junction, or to moor in a secret basin surrounded by traditional painted narrow boats, hidden in the heart of Birmingham are pleasures unique to BCN.

The heart of the BCN are the main lines from

Hidden in the centre of Birmingham and at the heart of the canal network, Gas Street Basin is the home of many traditionally painted narrow boats.

Birmingham's Gas Street Basin and its connections southwards with the Stratford and Worcester and Birmingham Canals to their northern connections near Wolverhampton with the Staffordshire and Worcestershire, and Shropshire Union Canals. Built between 1769 and the 1830s, these main lines encapsulate the history of the English canal system. Other BCN branches lead to the Grand Union Canal, for the south and London, to the Coventry Canal for the Midlands, and the Stourbridge Canal for the west and south-west. Notable among the parts of the BCN to have been lost is the **Wyrley and Essington Canal**. Closed during the 1950s and now only traceable in short sections, this meandering canal wound its way northwards through the Black Country to join the Coventry Canal near Lichfield.

Today the BCN is run down, occasionally depressing and often dirty. The only evidence of its frenetic life of 30 years ago is derelict and decaying buildings, overlooking gloomy waterways filled with rubbish. However, as industry retreats, so the BCN is slowly coming back to life, first as a secret and rarely disturbed natural wilderness, and second as a waterway network of great variety and interest, whose amenity potential is unrivalled in England. Redevelopment is under way, sometimes carried out with sympathy, as at Farmer's Bridge in Birmingham, and sometimes in a crude and mindless way as in the pointless removal of the Victorian warehouses that used to surround Gas Street Basin. There is a danger that the very qualities that make the BCN so unique will be swept away, but at the moment there is plenty of the eighteenth and nineteenth centuries still to be seen, a survival encouraged by the Black Country Museum at Dudley. Although they may not have a universal appeal, the unique attractions of the Birmingham Canal Navigations do deserve careful and sympathetic exploration.

The pretty flight of ten locks at Foxton on the Leicester Line of the Grand Union Canal, a busy and popular spot in the summer.

THE HEART OF ENGLAND

The river Trent flows through the heart of England in a great sweeping curve that effectively divides the Midlands from the North. With its connecting waterways the Trent forms a line from the Mersey to the Humber estuary. Although rarely explored, the Trent has considerable potential and interest. The surrounding landscape is predominantly farmland and woods, perhaps lacking in drama but attractively rural and typically English. The route of the Trent is full of history. Along its banks are Roman settlements, medieval castles and abbeys, fine Tudor and Georgian mansions, country villages, large cities and towns, the latter reflecting the rapid growth and expansion of the Victorian period. The Trent has been a river navigation for centuries, an important artery leading into the centre of England and today boats can still travel much of its length. The river is also the backbone of a waterway network that extends navigation in many directions, a radiating system built in the eighteenth century to bring the Industrial Revolution to some of the more remote corners of England. The landscape of the heart of England ranges from the bare hills, moorland and wooded valleys of the Peak District to the low-lying and desolate marshlands that flank the Trent's tidal estuary. In between are the miles of gently rolling farmlands that determine the character of the Midlands, a countryside whose subtle beauty deserves careful exploration. Rivers and canals offer the perfect way to discover and appreciate this style of countryside, and there is no shortage of pleasant, old-fashioned and meandering waterways to follow.

The same waterways reflect the impact of centuries of gradual development, making the history of the region and its sources of wealth easy to understand. In many areas the ridge and furrow of the medieval system of cultivation can still be seen, sometimes sharply divided by the line of an eighteenth-century canal. The sites of deserted medieval villages can be explored, as well as the great industrial cities of the Victorian period, whose growth was totally dependent upon the availability of navigable waterways. Stoke, Derby, Nottingham

The river Trent at Newark, a handsome town whose water-front is dominated by the ruins of its twelfth-century castle.

and Leicester are typical creations of eighteenth-century industrialisation, cities whose considerable wealth was founded upon their staple industries, pottery, coal, engineering, textiles, and their ability to export their products via the waterways. Stafford, Newark, Grantham and Macclesfield reflect an earlier period of history and an earlier form of wealth, based largely on agriculture and land management. The rivers played their part in the development of these more traditional towns and cities, and there are still plenty of traces of pre-industrial England to be discovered. The wealth of the region can be measured by the great number of fine houses and country estates, the tangible relics of a powerful, land-owning class whose considerable wealth inspired the building of river navigations and canals. Canals came early to this part of England, and many still survive in a virtually unchanged eighteenth-century state. Complete with great tunnels and other dramatic feats of engineering, they are fitting memorials to the ambition and ingenuity of their builders. The canal age was effectively launched in 1776 by the building of the Trent and Mersey Canal, the first of the grand trunk canals designed to link together the major waterways and industrial centres of England. From this waterway grew the vast network that made possible the Industrial Revolution, a turbulent period whose history can still be traced along the rivers and canals by the discerning explorer.

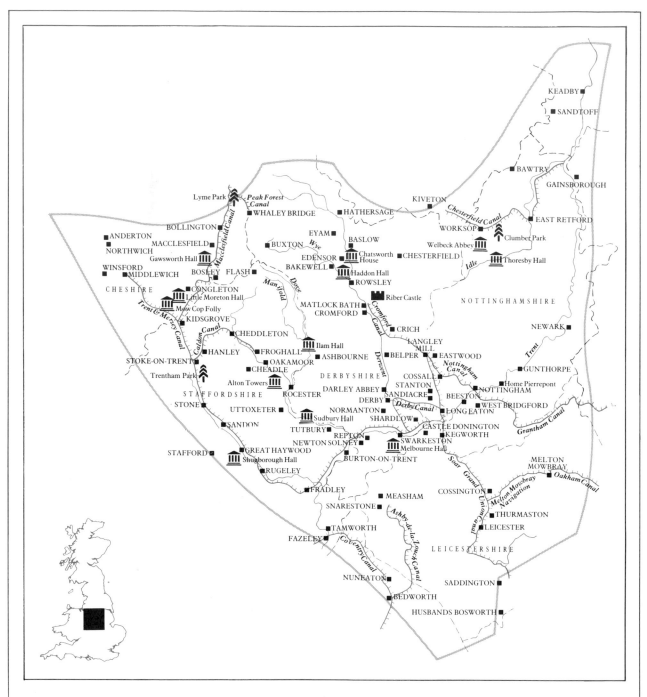

KEADBY

SANDTOFT

BAWTRY

GAINSBOROUGH

Lyme Park   *Peak Forest Canal*

WHALEY BRIDGE   HATHERSAGE

KIVETON

EAST RETFORD

Chesterfield Canal

WORKSOP

ANDERTON   BOLLINGTON

MACCLESFIELD

NORTHWICH

Gawsworth Hall

EYAM

BUXTON   BASLOW

Welbeck Abbey   Clumber Park

Wye

EDENSOR   Chatsworth House

WINSFORD

MIDDLEWICH   BOSLEY   FLASH

BAKEWELL   Haddon Hall

CHESTERFIELD   Thoresby Hall

Idle

CHESHIRE

ROWSLEY

Manifold   *Dove*

CONGLETON   Little Moreton Hall

Macclesfield Canal

NOTTINGHAMSHIRE

Mow Cop Folly

KIDSGROVE

Riber Castle

MATLOCK BATH

Trent & Mersey Canal   *Caldon Canal*

CROMFORD

Cromford Canal

NEWARK

CRICH

CHEDDLETON

LANGLEY MILL

HANLEY   FROGHALL

Ilam Hall

ASHBOURNE   BELPER   EASTWOOD

Trent

STOKE-ON-TRENT   OAKAMOOR

Derwent   *Nottingham Canal*

GUNTHORPE

Trentham Park   CHEADLE

COSSALL

Home Pierrepont

Alton Towers

DERBYSHIRE   STANTON

STAFFORDSHIRE   ROCESTER

DARLEY ABBEY   SANDIACRE

NOTTINGHAM

STONE

DERBY   BEESTON   WEST BRIDGFORD

Derby Canal   LONG EATON

UTTOXETER   NORMANTON

Grantham Canal

SANDON

Sudbury Hall   SHARDLOW

TUTBURY   CASTLE DONINGTON

REPTON   KEGWORTH

NEWTON SOLNEY   SWARKESTON

STAFFORD   GREAT HAYWOOD

Melbourne Hall   MELTON MOWBRAY

Shugborough Hall   BURTON-ON-TRENT

Soar   *Grand Union Canal*   Oakham Canal

RUGELEY

COSSINGTON

Melton Mowbray Navigation

FRADLEY   MEASHAM

THURMASTON

SNARESTONE   LEICESTER

TAMWORTH

Ashby-de-la-Tour-A Canal

LEICESTERSHIRE

FAZELEY   *Coventry Canal*

SADDINGTON

NUNEATON

BEDWORTH

HUSBANDS BOSWORTH

The Trent

The course of the **Trent**, which runs through Staffordshire and Nottinghamshire, falls into two distinct parts. The first stretch, from the Trent's source near Stoke to Shardlow, midway between Derby and Nottingham, winds along a varied but generally unexciting route, lined by farmland, willow trees and the occasional village or town. Leaving behind the potbanks and housing estates of Stoke, the Trent flows through Trentham Park towards Stone, where it is joined by the Trent and Mersey Canal. River and canal then share the same valley, passing Stafford and Rugeley, with Cannock Chase and Shugborough Hall to the south. From Rugeley to Burton the Trent follows its own route, although the canal is never far away. Burton, the brewing capital of England, grew up around the Trent which, during the seventeenth and eighteenth centuries, was navigable through the town. East of Burton the river moves briefly into Derbyshire.

The importance of the lower reaches of the river Trent for commercial traffic is underlined by this industrial scene at Torksey.

After Repton it flows under the medieval bridge near Swarkestone, passes Melbourne Hall and the racing car museum at Castle Donnington, and then it reaches Shardlow, the start of the Trent Navigation and the eastern terminus of the Trent and Mersey Canal.

The second part of the Trent, from Shardlow to Trent Falls where it joins the Humber estuary, is very different. Navigable for several centuries, and steadily improved since the eighteenth century, the Trent quickly becomes a major waterway, with 13 large locks to facilitate the passage of boats along the 95 mile route to the estuary. At first it follows an industrial course through Long Eaton, Beeston and Nottingham, curving round the castle mound and passing the Nottingham Canal Museum housed in an old Trent warehouse, but after the junctions with the Derwent and Soar, the river moves into a more open landscape filled with the lakes formed from old gravel workings. At Holme Pierrepoint, one of these has become an international rowing and water sports centre. The Trent is not a pretty river, but there is plenty to see in the surrounding towns and villages and a variety of boats use the river. Some commercial traffic still survives, notably barges carrying stone, gravel and oil. Gunthorpe, the site of a Roman river crossing, is a good place to watch the boats. Small and rather isolated villages follow the river to Newark, a handsome town with a twelfth-century castle, a fine parish church with a huge spire, some good Georgian buildings and an interesting variety of traditional shops. North of Newark, at Cromwell Lock, the Trent becomes tidal and assumes the character of a wide, fast-flowing, powerful waterway requiring skill and experience to navigate. Its high banks and the low-lying landscape of the region limit the views of the river, which follows a winding course to Gainsborough, passing at Torksey the mouth of the Fossdyke Navigation which leads to Lincoln. Gainsborough is a pleasant small town that retains the atmosphere of a nineteenth-century inland port and is still busy with a variety of commercial traffic. North of Gainsborough the size and power of the Trent steadily increases and it becomes a daunting sight as it approaches the estuary. At West Stockwith the river passes the junction with the Chesterfield Canal and the river **Idle** (navigable since the seventeenth century to Bawtry but visually unexciting) and then the surroundings become more industrial as the Trent turns into a mighty seaway, navigable only by large ships. At Keadby is the junction with the Keadby and Stainforth Canal, a link with the canals and navigable rivers of South Yorkshire.

The river Derwent adds the all-important dimension of water to the magnificent gardens of Chatsworth House.

The Chesterfield Canal

Opened in 1777 to link Chesterfield with the river Trent, this 46 mile canal had 65 locks and one major tunnel. Despite its isolation from the rest of the English canal network, the **Chesterfield Canal** was a commercial success throughout the nineteenth century and until 1908 when the tunnel at Norwood collapsed. This disaster effectively sealed off the canal west of Worksop, leaving the western section to Chesterfield to fall into gradual decay, while the eastern section, from Worksop to the Trent at West Stockwith continued to carry commercial traffic until the 1950s. Since then the canal has been used increasingly by pleasure boats as it offers a rural solitude unequalled by many other English waterways. From West Stockwith to East Retford the canal follows a winding and rural route, flanked by small and pretty villages and then beyond Retford it enters an area known as the Dukeries, a wooded, rolling landscape richly decorated with country houses. Nearby are Clumber Park, Welbeck Abbey, Thoresby Hall, houses built on the fortunes made from the Nottinghamshire coal fields in the eighteenth century. The coal fields made the canal successful but brought about the collapse through subsidence of the tunnel. The present terminus of the canal is just beyond some impressive warehouses in Worksop, but the route to Chesterfield can easily be explored, to discover lock flights, the remains of the great Norwood Tunnel, which is over 3000 yards long, and the remote, wooded stretches of canal between Kiveton and Worksop.

The rivers of the Peaks

The Peak District, an oasis of wild and rugged country surrounded on most sides by urban and industrial development, can be enjoyably explored by following its rivers. The **Derwent**, the **Dove**, the **Manifold** and the **Churnet** all flow rapidly from their sources high in the Peaks towards the Trent, their routes marked by steep valleys, remote farms and villages and frequently spectacular scenery. The greatest of these rivers is the Derwent, which rises in the inaccessible moorlands above the Derwent Reservoir, overlooked by Kinder Scout, the Dark Peak and other hills that reach towards the 2000 foot level. From the reservoir southwards, the Derwent is accompanied by minor roads and so exploration is easy. With Sheffield only a few miles to the east, the Derwent is surprisingly undeveloped, its natural privacy protected by the inhospitable nature of the terrain. Typical Peak villages, Hathersage, Eyam and Baslow, cluster round the river in its valley, and then south of Baslow the landscape becomes more open as the Derwent flows through Edensor and the grounds of Chatsworth House. Water is an essential feature of Chatsworth, its classical bridges, mighty cascade and huge fountain all make a major contribution to the park. Still in a valley but now a substantial waterway, the Derwent continues to Matlock and Matlock Bath, dominated on the west by hills rich in prehistoric remains, and on the east by the nineteenth-century Riber Castle and its wildlife park, and the wooded Heights of Abraham. At Cromford there are the great mills established by the Arkwrights in the eighteenth century, mills that originally were dependent upon the Derwent for their power. This is another region where the remains of the Industrial Revolution can be explored and wondered at, the mills, the lead mines, the pumping engines, the ironworks, early railways and the Cromford Canal. To the south is another opportunity to discover a vanished aspect of English life, the Crich tramway museum. The Derwent flows through Belper, another mill town, before entering the outskirts of Derby, passing Darley Abbey. Derby does not make much of its river, and there is little to show that the Derwent was once a busy navigation that connected Derby to the Trent. The navigation was abandoned in the late eighteenth century, however, after the opening of the Derby Canal, which offered a more direct route. South of Derby the Derwent flows through unexciting country to join the Trent near Shardlow.

The Derwent has one tributary of importance, the river **Wye**. This rises north of Buxton, passes attractively through the town, and then enters the

The pleasures of tramway travel can be re-experienced at the Crich Tram Museum, a mecca for tram enthusiasts and time travellers.

Dovedale in Derbyshire, the most popular of the Peak District's limestone dales. It offers some remarkably varied scenery and strange rock formations.

dramatic valley of Miller's Dale. Inevitably popular among visitors to the Peaks, the Wye continues on through Bakewell to its junction with the Derwent at Rowsley. Old mills mark its route, while Haddon Hall overlooks its valley. The Dove and the Manifold are Peak rivers of similar attraction and interest, flowing along parallel valleys before joining near Ilam Hall. The Dove, which forms the border between Derbyshire and Staffordshire, rises near Flash, England's highest village. Its valley is remote, undeveloped and accessible only on foot for most of its route, but its splendid scenery, particularly through Dovedale, is well worth the effort. The Manifold is accessible from minor roads, and its valley is more dramatic: caves, swallow holes and precipices mark its route. South of its junction with the Manifold, the Dove is a more significant river, passing west of Ashbourne before entering the thickly wooded valley known as the Staffordshire Rhinelands. Nearby is Alton Towers, the only pleasure park in England that attempts to rival Disneyland. Leaving behind the attractions of the Peaks, the Dove becomes a gentle river. It winds its

way through farmland, round the outskirts of Uttoxeter, through the grounds of seventeenth-century Sudbury Hall, past Tutbury with its castle and then to the junction with the Trent near Newton Solney.

The fourth of the Peak District rivers is the Churnet, which like the Manifold has its own distinct character despite being a tributary of the Dove. Rising north of Rudyard Reservoir, the river winds its way round Leek and then enters the attractively wooded Churnet valley, along a route which it shares with the Caldon Canal. At Cheddleton are the preserved mills where the flints for the pottery industry were ground by water power and nearby is the Churnet Valley Steam Railway. Froghall and Oakamoor are quiet towns in the Staffordshire moorlands, a region of rural isolation still unaffected by the proximity of Stoke-on-Trent and the Potteries, while to the south is Cheadle, with its magnificent church designed by Pugin. The Churnet then passes through the grounds of Alton Towers, flanked by wooded hills and, shortly after Rocester, it joins the Dove.

The canals of Derbyshire and Nottingham

The importance of the river Trent as a navigation inevitably encouraged the building of a number of canals during the eighteenth century designed to connect the new centres of industry to the river, and thus to the developing national waterway network. Some of these canals are still in existence today, but others have largely disappeared. Besides paying little attention to its river, the Derwent, Derby is one of the few major cities and centres of industry to have lost completely its connection with the canal system. Throughout the eighteenth century barges were able to reach Derby along the Derwent, but in 1796 this journey was made easier by the opening of the **Derby Canal** which ran from the Erewash Canal westwards to Derby, through the city, and then south to join the Trent and Mersey Canal at Swarkestone. Commercial traffic continued to use the canal until the early 1960s and it was closed finally in 1964. The owners of the Derby Canal somehow managed to escape the nationalisation of

Bottle ovens are now a rare canalside sight in Staffordshire. These examples are preserved at the Gladstone Pottery Museum, Stoke-on-Trent.

the rest of the canal system and so were able to sell off much of the course of the canal for development during the 1960s. As a result, there is very little of the Derby Canal to be seen today, and even the Holmes Aqueduct, the first cast iron aqueduct in the world, has been destroyed.

The earliest canal in this area is the **Erewash**. Opened in 1779 from Sawley on the river Trent to Langley Mill, 12 miles to the north, the Erewash was the backbone of a small waterway network serving the coal mines and ironworks of the region. Ironically it is the only part to remain navigable, all the later developments having been closed. Its route is not particularly exciting as it mostly runs through a mixture of urban and industrial surroundings, but there are some features of interest, notably the canalside lace mills at Long Eaton and Sandiacre (where there are also the remains of the junction with the Derby Canal), the Stanton ironworks, Eastwood, which was the birthplace of D.H. Lawrence and the recently restored Langley Mill Basin. A century ago Langley Mill was a bustling canal centre, for two other waterways came together here.

The first of these waterways, and by far the most interesting, is the **Cromford Canal**. Opened in 1794, the canal ran from Cromford on the river Derwent, the home of the Arkwright cotton mills, to Langley Mill where it joined the Erewash. It was a dramatic canal, cut along the steep Derwent valley and it had a number of unusual features, including several aqueducts and the 3000 yard Butterley Tunnel. Coal, stone, lead and iron were the staple cargoes on the canal, which thrived until 1900, when the tunnel collapsed, effectively cutting the canal into two parts. Decay was fairly rapid, particularly on the section between Cromford and Butterley, and the canal was quickly abandoned. However, that is not the end of the story, for part of the upper section of the Cromford Canal has recently been restored, largely by volunteers. Today, Cromford is once again a centre of canal activity. The extraordinary ingenuity that inspired the Industrial Revolution can be explored at Cromford Wharf, with its original warehouses; High Peak Wharf, where the trucks of the Cromford and High Peak Railway started their strange journey aross the hills and moorlands of Derbyshire to Whaley Bridge, by being hauled up the first of a series of inclined planes; the restored Leawood steam pumping engine, which lifts water from the Derwent into the canal; the handsome Wigwell aqueduct. There are boat trips along the restored section, which eventually will reach Whatstandwell and Ambergate. From Ambergate southwards much of the canal has disappeared including most of the locks and the Pinxton branch, but the daunting portals of Butterley Tunnel are still to be seen.

The second canal to join the Erewash at Langley Mill is the **Nottingham**. This ran from the Trent, near Trent Bridge, to Langley Mill, passing through the centre of Nottingham, and it was designed to allow traffic from the Cromford Canal, particularly coal and iron, a direct route to Nottingham and the Trent. Commercial traffic continued until the 1920s and the canal was closed in 1937. Much of the Nottingham Canal has now vanished but there are the remains of locks at Wollaton and an aqueduct at Cossall, and between Trowell and Cossall a stretch survives which still holds water. The Nottingham Canal deserves a better fate than the total obliteration that has been the lot of its neighbour, the Derby Canal, for although not particularly scenic, its remains are a tangible witness to the industrial development of the region in the late eighteenth century.

The Trent and Mersey Canal and its branches

One of the greatest of the English canals as regards both length and historical importance is the **Trent and Mersey** or, as it was called at first, the Grand Trunk Canal. It was planned originally as part of the great cross of canals designed to link the Mersey, the Trent, the Thames and the Severn, and it was the first of the waterways to link together the two coasts of England. James Brindley was the engineer and work started in 1766 on the 93 mile route. Among the promoters of the canal was the potter Josiah Wedgwood, who realised that the future development of the pottery industry in Staffordshire was dependent upon an improved system of transport between the factories in Stoke and the docks at Liverpool. As a result his new factory, built at Etruria during the 1760s, was situated beside the canal. The Trent and Mersey runs from Preston Brook, south of Runcorn, where it leaves the Bridgewater Canal to Shardlow in Derbyshire where it joins the Trent Navigation. Despite its industrial associations, the Trent and Mersey is a predominantly rural canal that enjoys a continously changing landscape. It was a difficult and expensive canal to build and its completion was delayed by a number of engineering problems, notably the digging of the great tunnel through Harecastle Hill, north of Stoke. When it was finally fully opened in 1777, the canal was an immediate success and was directly responsible for the rapid growth of a number of towns and cities along its banks, in particular Northwich with its salt industry, Stoke and its potteries and Burton with its breweries. Commerical traffic continued to use the Trent and Mersey Canal until the 1960s, with pottery materials as the staple cargoes right up to the end.

The Trent and Mersey is an interesting and varied canal to explore, with 76 wide and narrow locks, including some in pairs, several tunnels and aqueducts and a variety of attractive bridges. At first the canal follows the wooded valley of the river Weaver but at a higher level so there are good views of the river and its shipping. At Anderton is another wonder of the waterways, the unique vertical lift that transports boats between the Trent and Mersey and the Weaver below. This extraordinary piece of Victorian engineering, built in 1875, consists of two great iron water tanks, in which boats are raised or lowered. From here onwards the landscape is dominated by the salt industry, with mines at Marston, Northwich and Middlewich. Continual subsidence has caused the canal to be raised high above the surrounding landscape. At Middlewich there is a junction with the branch from the Shropshire Union Canal to the west, and then a long flight of locks carries the canal round Sandbach and up to Kidsgrove and the Harecastle Tunnel. Just north of

Hazelhurst Locks on the Caldon Canal, which was reopened in 1974 after years of disuse. The branch to Leek leaves the main line here.

The vertical lift at Anderton, one of the wonders of the waterways. It transports boats from the Trent and Mersey Canal to the Weaver below.

the tunnel is the junction with the Macclesfield Canal, which leaves the Trent and Mersey on the south (crossing it on a fly-over aqueduct) after which the canal proceeds north towards Manchester. There are actually two Harecastle Tunnels. James Brindley's original bore was closed by subsidence over 50 years ago, leaving Thomas Telford's tunnel of 1827 as the only way through the hill. The mouths of the two tunnels still stand side by side and the passage through is a daunting experience as the roof is only six feet above the waterline. After Harecastle, the canal passes through the centre of Stoke, although most of the traditional canalside potteries with their characteristic bottle-shaped ovens have disappeared. Near the centre of Hanley, one of Arnold Bennett's 'Five Towns', is the

The northern portal of the narrow Harecastle Tunnel on the Trent and Mersey Canal

at Kidsgrove in Staffordshire. The canal water is stained red by underground mineral deposits.

Froghall Basin, the terminus of the Caldon Canal. The Basin features a warehouse and some giant lime kilns and is approached through a low, narrow tunnel.

junction with the Caldon Canal, while a number of smaller canals and branches, now mostly derelict or vanished, used to connect various pottery factories to the Trent and Mersey. All the great names of the industry – Wedgwood, Minton, Spode, Doulton – owed their success to the Trent and Mersey and its branches during the eighteenth and nineteenth centuries. South of Stoke the canal runs through pleasant rolling farmland, passing Stone, Sandon, Great Haywood and the junction with the Staffordshire and Worcestershire Canal, Shugborough Hall and Rugeley, a route shared with the river Trent. At Fradley there is a junction with the Coventry Canal and then, after a rural and rather remote section, the canal passes through Burton, following the line of the Roman Ryknild Street. An aqueduct carries the canal over the river Dove and at Swarkestone is the short arm that used to lead to the Derby Canal. The final village before the junction

with the Trent is Shardlow, a classic canal settlement complete with elegant warehouses and original wharves, richly ornamented with eighteenth century architectural details.

Although technically a branch of the Trent and Mersey, the **Caldon Canal** has its own distinctive personality. Seventeen miles long, the Caldon runs from the Trent and Mersey Canal at Etruria to Froghall. It was opened in 1779 and a short branch to Leek was added in 1802, followed by an extension from Froghall to Uttoxeter in 1811. The canal was built primarily for the transport of stone, which was brought from quarries at Cauldon Low to Froghall by a series of tramways. Never officially closed (with the exception of the Uttoxeter section which was converted into a railway in 1847), the Caldon survived in a rather derelict state until the early 1970s, when an ambitious restoration programme was launched. It was reopened to Froghall in 1974, along with much of the Leek branch. The Caldon is a canal of remarkable contrasts. It leaves Hanley and the potteries through a nineteenth-century industrial landscape, and then is suddenly in the heart of the countryside, surrounded by woods and moors, and overlooked by the hills of the Peak District. Between Cheddleton and Froghall the canal follows closely the valley of the Churnet, one of the most remote and beautiful stretches of canal in the country. At Froghall Basin there is a tiny tunnel, some warehouses and, almost buried in the woods, a dramatic series of lime kilns overlooking a sensitively planned picnic area.

The Macclesfield and Peak Forest Canals

The **Peak Forest** and **Macclesfield** Canals formed an important waterway route into Manchester from the south. The Peak Forest was the first to be built, running from Dukinfield on the Ashton Canal east of Manchester to Buxsworth Basin and Whaley Bridge, 14 miles to the south-east. Stone quarries, connected to Buxsworth by tramways, supplied most of the traffic and as late as the 1880s up to 30 boats a day were leaving the basin. However, this traffic died during the 1920s after which the Peak Forest began a slow decline and was impassable by the 1960s. Closure seemed inevitable but an ambitious restoration scheme using volunteer labour saved the canal, and brought about its reopening in 1974.

Once away from Manchester's suburbs, the Peak Forest quickly becomes a canal of great attraction, the section from Marple to Whaley Bridge is among the most dramatic in the country. At Marple there are 2 short tunnels and a magnificent stone aqueduct over the river Goyt with 3 huge arches that is over 100 feet high. A flight of 16 locks lifts the canal to an embankment built on to the side of the Goyt valley.

The Marple Aqueduct, a handsome three-arched stone aqueduct that carries the Peak Forest Canal over the river Goyt above a dramatic wooded valley.

Thickly wooded, with splendid views over the valley, this embankment continues to Whaley Bridge, passing the junction with the Macclesfield Canal. At Whaley Bridge there is a terminus basin with an original stone warehouse in which goods were transhipped from boats to railway waggons for the journey across the Peaks to Cromford, along the Cromford and High Peak Railway. Although long closed, the course of the railway can be followed on foot and the remains of its inclined planes explored. This is a worth while exercise as it clarifies the importance of this vital railway link between the canals of Derbyshire and the Midlands, and the canals of Manchester and the north-west.

Built comparatively late, the Macclesfield Canal formed an important link between the Peak Forest and the canals of Manchester and the Trent and Mersey. It is a handsome canal, built largely on the

One of the Macclesfield Canal's characteristic curving bridges, designed so that the towing horse could cross the canal without being unhitched.

The timber-framed extravagance of Little Moreton Hall, near the Macclesfield Canal.

The Weaver

One of the lesser known of the English river navigations, the **Weaver** has continued to enjoy a busy commercial life, the result of steady improvement and development since 1732 when it was first made accessible to barges. The secret of the Weaver's success is the salt trade, for salt, and more recently chemicals, have always been the staple traffic. The Weaver rises south of Nantwich and then flows through a quiet landscape of farmland to Winsford, where the navigation begins. Large locks and swing bridges mean that boats up to 130 feet long and 35 feet wide can use the Weaver, and so there is always a variety of shipping to be seen, particularly in the section between Northwich and the Mersey. Despite its importance as a commercial waterway, the Weaver is still an attractive river. Some of the locks, at Dutton, Saltersford and Vale Royal, are set in attractive wooded countryside, still pleasantly remote despite the proximity of industry. Between Northwich and Frodsham the Trent and Mersey Canal flows beside the Weaver, but at a higher level, and the only link is via the vertical boat lift at Anderton. For the last four miles the Weaver is tidal, and so it has been bypassed by a canalised section that takes boats to Weston Point Docks, where there is a connection with the Manchester Ship Canal and the Mersey estuary.

The canals and rivers of the east Midlands

The east Midlands are served by two canal and river networks that enable boats to travel from the waterways of the south to those of the north. The first, formed largely by the **Coventry Canal**, connects the Grand Union and Oxford Canals from London and the south to the Trent and Mersey Canal and to the Birmingham Canals. The second, formed by a branch of the Grand Union and the river Soar, leads to the river Trent, and thus to the north-east.

Opened in 1790, the Coventry Canal runs from its terminus basin in the heart of Coventry to Fradley, where it joins the Trent and Mersey Canal. There are two other vital junctions on its route, first with the Oxford Canal at Hawkesbury, and second at Fazeley with the Birmingham and Fazeley Canal which leads to the Birmingham Canal Navigations. Traditionally a coal canal, the Coventry remained in commercial use until the 1960s, and coal mines, both active and long abandoned, still accompany the canal for much of its route. Despite this, it is not

500 foot contour and its route is marked by impressive embankments, notably at High Lane, Bollington and Bosley, which offer exciting views over the wooded river valleys and rolling farmland that mark the edge of the Cheshire Plain. Another intriguing feature of the canal are the elegant stone bridges that carry the towpath from one side to the other, designed so that the horse could be led across without having to be detached from the barge. The route of the canal is from Marple on the Peak Forest to Hall Green, near Harecastle on the Trent and Mersey, a route through Bollington, Macclesfield and Congelton. Besides the attractions of its landscape and architecture, the Macclesfield Canal also enjoys a variety of fine houses near its banks, including Lyme Park, Gawsworth Hall and the black-and-white splendour of Little Moreton Hall. There are several aqueducts and a flight of 12 locks in remote country near Bosley. Between Congelton and the junction with the Trent and Mersey, the canal is dominated by two great hills, The Cloud, and Mow Cop, the latter crowned by an eighteenth-century folly in the form of a ruined castle. Handsome, elegant, and built through a dramatic landscape, the Macclesfield Canal represents one of the final attempts by the great engineer Thomas Telford to hold the railway age at bay.

*The rolling farmlands of the Midlands provide an attractive setting for the Coventry Canal,
seen here near Polesworth.*

unattractive, with long stretches running through woods and farmlands, passing a number of places of interest and some fine canal architecture. In Coventry itself the basin still has many of its original buildings and is well placed near the cathedral. Arriving in the city by boat is certainly more enjoyable than by car. Hawkesbury Junction is a traditional canal centre, retaining much of its nineteenth-century atmosphere and there is usually a good selection of working narrow boats to be seen. Leaving Hawkesbury the canal passes through largely urban surroundings to Nuneaton, and from there to Tamworth its route is enjoyably rural, with an attractive flight of locks through Atherstone. At Tamworth there is an aqueduct over the river Tame, overlooked by the castle. From Tamworth to Fradley the canal runs through a wooded landscape, its route bypassing Lichfield and marked only by small villages, including Huddlesford where there was a junction with the abandoned Wyrley and Essington Canal. Fradley Junction itself is a popular spot with an attractive pub and canal buildings.

There is one other aspect of the Coventry Canal that should not be overlooked, namely the two branch canals that join it. The first of these is the **Ashby-de-la-Zouch Canal**, which leaves the Coventry near Bedworth and then meanders northwards for 22 miles before coming to a stop in the middle of a field near Snarestone. Opened in 1794 to serve the coalfields of Leicestershire and south Derbyshire, the canal was remarkably successful and remained in commerical use until the late 1960s. Despite its name it never reached Ashby, although it used to have a more substantial terminus in Measham, the top few miles having been closed through subsidence. Today the Ashby is a totally rural and little used waterway, winding its way through the pretty villages of Leicestershire. It is rarely visited by boats because it does not really go anywhere, apart from passing the site of the Battle of Bosworth Field. River-like in quality, the Ashby is an idyll of peace and quiet in an overdeveloped region. Rather more unusual but largely vanished are the **Newdigate Canals**, a network of private canals built in and around his Arbury Hall estate by Sir Roger Newdigate between 1769 and 1796. The entrance to this network can still be seen north of Hawkesbury Junction but little else remains of the system that totalled over 8 miles and included 13 locks. Constructd to serve the coal mines around Arbury Hall and used for transport around the estate, the network declined during the early nineteenth century until eventually only the access waterway from the Coventry Canal remained in use. The remains of some locks and short sections of canal still survive in the estate and can be tracked down by energetic visitors to Arbury Hall.

The **Leicester** arm of the Grand Union Canal leaves the main line at Norton Junction, an area now overshadowed by the Watford Gap service station on the MI motorway. A through route from the Grand union to the Trent, the Leicester arm was formed by the amalgamation of several small canal and river navigations, mostly dating from the late eighteenth century. A predominantly rural and rather meandering waterway, its commercial success was limited by the mixture of wide and narrow locks, which effectively prevented the larger boats on the Trent from penetrating further south than Leicester. Such restrictions do not affect pleasure boating and today the route is heavily used, particularly the remarkably attractive and remote section from Watford Gap to Leicester. There are few villages, but the canal enjoys good views over a rolling and wooded landscape, and there are a number of interesting tunnels, at Crick, Husbands Bosworth and Saddington. However the main feature of this section is the flight of ten locks at Foxton, an unusually pretty flight busy with boats during the summer. Nearby can be seen the remains of the extraordinary inclined plane, which raised and lowered fully laden narrow boats in huge tanks on rails. Powered by steam, and designed to replace the locks, this grandiose machine was only used between 1900 and about 1912. At the bottom of the Foxton flight is the branch leading to Market Harborough. The main line continues its wandering course towards Leicester, remaining rural and remote until the outskirts of Leicester itself. South of the city the canal joins the river **Soar Navigation**, which passes through the centre of Leicester in a handsome cutting. Unlike many Midlands towns

Despite its rural surroundings, the river Weaver in Cheshire is still a busy commercial waterway used by coasters carrying chemicals and salt.

Closed to boats in the 1950s, the Grantham Canal is typical of England's forgotten waterways and is one of the easiest to explore.

Originally built to carry coal, the rural Ashby Canal now follows a winding river-like course through miles of quiet Midland scenery.

and cities, Leicester seems proud of its waterway. North of Leicester the Soar widens into a major river, with fast-flowing weirs and large locks. The surroundings are still rural but a number of gravel pits accompany the river through a more open landscape. There are some pretty villages, Thurmaston, Barrow, Normanton and Kegworth, but as a rule buildings tend to stay well back from the river banks for the Soar is liable to sudden flooding. The Soar bypasses Loughborough (not an attractive town in any case) but nearby is the steam-hauled Great Central Railway. The Soar joins the Trent at Red Hill lock, in the shadow of the huge power station at Radcliffe. Although there is very little commercial traffic, the Soar is a popular waterway, much used by pleasure boats and easy to explore by road. Less known is its connection with the former **Melton Mowbray Navigation** and the **Oakham Canal**, a largely forgotten waterway whose junction with the Soar at Cossington can still be seen. This obscure waterway was opened in the late eighteenth century when the river Wreake was made navigable to Melton Mowbray. A few years later, in 1803, the 15 mile long Oakham Canal was built from Melton Mowbray to Oakham, making a total waterway of over 30 miles long with 31 locks. This rather unlikely system traded successfully for a number of years but the Oakham Canal was closed in 1846 and the Melton Mowbray Navigation followed it 30 years later. Relatively easy to explore today, the waterway offers a number of attractions, including an attractively rural route, the remains of locks and wharves, some canal buildings and not least a golden opportunity to appreciate one of the more extraordinary schemes devised by eighteenth-century canal enthusiasts.

The Grantham Canal

One of the simplest of England's lost waterways to explore is the **Grantham Canal**. Opened in 1797, this 33 mile long rural canal ran from Grantham to the Trent at West Bridgford, its route wandering across the Vale of Belvoir. Boats continued to use the canal regularly until the 1930s and it was not totally closed until the 1950s, when locks were removed and bridges lowered. Since then, little has happened and the canal remains largely intact. It has water throughout much of its length and is easily explored on foot or by car with the help of a good map. More like a river than a canal, the Grantham has been the subject of various restoration schemes, but new coal mines planned for the Vale of Belvoir make any eventual reopening to boats more unlikely.

The parish church at Thaxted in Essex, one of the many attractions along the course of the river Chelmer.

THE EAST

The character of the eastern counties has been largely determined by the rivers and waterways. It has always been an isolated and self-contained area, and for centuries was dependent upon waterways for its links with the rest of the country. Roads were few and far between in a landscape composed of low-lying marshland prone to flooding and incursions by the sea. The marshland created natural barriers to isolate the region, an isolation encouraged by the great forests that used to cover the western part of Essex. Over the centuries the coast, with its miles of mud flats and tidal estuaries, has been created by the constant battle between sea and land. In some areas the sea has been driven back, in others it is still eroding the land. From Roman times the rivers have been exploited, for transport and as a means of draining the land and keeping the water level under control.

The rivers of eastern England fall into three distinct groups. First, there are the rural rivers of Essex and Suffolk, flowing through a rolling landscape of arable land and woodland, linking small villages and dominated by the spires and towers of innumerable churches. These rivers reach the sea via huge tidal estuaries surrounded by mud flats, the home for generations of fishermen and sailors. Individual and independent, these rivers are at the heart of the English tradition of landscape painting. Second, there are the rivers and Broads of Norfolk, a complex network of partly man-made and partly natural waterways that connect many of the major towns and villages of the region, and serve both for transport and drainage. Now, they are one of the great recreational waterways of England, frequently overcrowded, but still able to maintain their own distinctive quality. Third, there are the rivers and waterways of the Fens, a huge interconnected network of major rivers and drainage channels grouped round the Wash. This region was first tamed by the Dutch engineers of the seventeenth

The church at Hemingford Grey on the Great Ouse. The church spire was blown into the river during a storm in 1741.

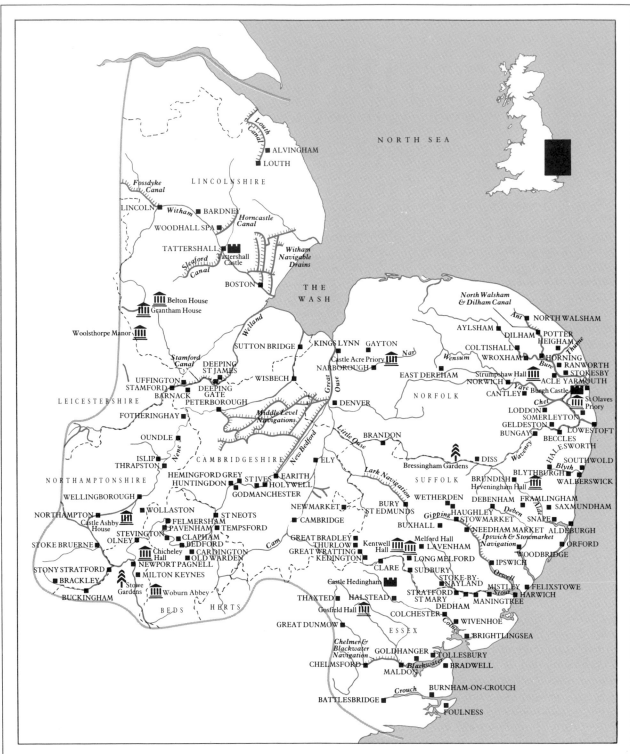

century, whose legacy is miles of navigable waterway, some of the richest farmland in England, and the remains of hundreds of windmills. Until the early nineteenth century and the coming of steam, over 700 wind-driven pumps kept the water levels under control. This area stretches from the flat landscape of the Fens to the Lincolnshire Wolds. There are many reminders of the dramatic changes that the region has witnessed since the seventeenth century. Many towns and villages are built on low hills, land that would formerly have been islands in a great sea of impenetrable marsh. The remains of a

Burnham, on the tidal river Crouch in Essex. One of England's premier sailing centres, it attracts yachtsmen from far and wide.

number of abbeys and castles underline the feeling of isolation and independence that characterises the region. Yet, this is not a poor area. It benefited from successive waves of colonisation and there are traces to be found of the Romans, the Saxons and the Danes. More recently, the influx of Flemish weavers, builders and brewers helped to create the enormous wealth associated with the wool industry, from which came the magnificent churches to be found all over eastern England, including the four great cathedrals of Peterborough, Norwich, Ely and Lincoln. The final impetus came from Holland, which supplied the men who built the ports, controlled the rivers and drained the land. All these groups left their mark, on the buildings, on the roads, on the names, creating layer by layer a part of England that is distinctive, individual and highly cultured. Centuries of development have still left many areas of natural wilderness. The mud flats of the estuaries and parts of the Broads and Fens contain a great wealth of birds, plants and insects, which can be enjoyed by the visitor in addition to the variety of rivers, the changing landscape, and the special quality of light that can only be found in a region where the horizons are generally low.

The Essex sailing rivers

The Essex coastline has been formed by the continual battle between the sea and the land and so much of it is composed of marshland and mud flats broken up by the long fingers of tidal river estuaries. It is an isolated coastline, rich in bird life, with many areas only accessible by boat. For centuries the rivers have been used for trade, for fishing and for smuggling, allowing a number of inland towns and villages to develop as ports of some significance. Today the emphasis has shifted and while there are probably more boats than ever using the Essex rivers, most are there for leisure purposes. The river **Crouch**, which is navigable for 17 miles from Battlesbridge to its estuary at Foulness, is now one of the great sailing rivers of England. There are many yacht clubs, including some of the grandest in the country in and around Burnham-on-Crouch, which is a good place to watch sailing both amateur and professional. An even greater variety of boats can be seen on the river **Blackwater**, a wide tidal estuary navigable as far as Maldon. This is still a river for fishing boats, oyster beds and the traditional Thames sailing barges. Many of the villages, Tollesbury, Bradwell and Goldhanger for example, were formerly centres of local trade, and reveal traditions that are sometimes centuries old. There is a Saxon chapel near Bradwell, while Maldon, still a busy sailing centre, has a range of interesting buildings dating from the sixteenth to the nineteenth centuries. Maldon is also the terminus of the **Chelmer and Blackwater Navigation**, a canal opened in 1797 to link Chelmsford with the sea. This quiet and pastoral waterway remained in commercial use until the 1970s. It was used mainly by timber barges and now is largely recreational. The river **Chelmer** continues beyond Chelmsford, a rural stream flowing through the cornfields of Essex and linking towns such as Great Dunmow and Thaxted, where good examples of timber-framed and brick-built Tudor houses can be seen, as well as fine churches, Georgian streets and a preserved windmill.

To the north of Mersea Island, famous for its oyster beds, is the estuary of the river **Colne**. This rises near the Suffolk border, flows through Castle Hedingham, overlooked by the keep of the Norman castle, and the traditional market town of Halstead, then, passing near Gosfield Hall, the finest Tudor mansion in the county, it continues to the Roman town of Colchester, the scene of Boadicea's triumphs over the Roman army. Today, Colchester, with its Roman walls, Norman castle and priory,

and range of buildings of later periods, is still a busy inland port. Small coasters sail up the Colne to the town quays on the tide, even though at low water the river is no more than a trickle. Between Colchester and the sea the river passes Wivenhoe, formerly a centre of boat building, and the fishing village of Brightlingsea.

The Stour

The **Stour**, which for much of its course marks the boundary between Essex and Suffolk, is a river firmly established in the English consciousness by the paintings of Constable. The Stour valley and its villages of Flatford, Dedham, Stratford St Mary and Stoke-by-Nayland must be one of the most familiar stretches of waterway in England, and yet there is far more to the Stour than this short section. In 1821 Constable wrote: 'I associate my careless boyhood to all that lies on the banks of the Stour.' Today, an exploration of the river can be just as rewarding, even if some of the banks have changed.

The Stour rises in the low-lying farmlands south of Newmarket and flows through villages of unchanged and unchanging quality, Great Bradley, Thurlow, Great Wratting, Kedington, on to Clare, a handsome town dominated by the ruined castle and priory. Nearby the river is flanked by vineyards, a revival of an industry that flourished in the Roman

Wivenhoe in Essex. An old fishing village on the tidal river Colne, it is now enjoying a new lease of life as a holiday and leisure centre.

Constable's The Leaping Horse, *a typical nineteenth-century Stour valley scene.*

period. This is an area rich in fine houses, and even finer churches, the latter reflecting the former wealth of the wool merchants. Traditional timber framed buildings can be seen in Long Melford and Lavenham, while Kentwell Hall and Melford Hall reflect the glories of Elizabethan brickwork. Sudbury, an old fashioned market town, is the birthplace of Gainsborough, and establishes the Stour as an artist's river. Sudbury also used to be the limit of navigation. From about 1705 to 1916 the Stour was used commercially between Sudbury and the estuary, and traffic survived on the lower reaches until the 1930s. Today, some of the Stour locks are being restored, and so boats may well one day return to Sudbury, to bring back to life the old quay warehouse. Many of the characteristic features of the Stour navigation are preserved in Constable's paintings, notably the meandering and rural course of the river, its distinctive locks, the locally built barges, the barge horses leaping the fences built across the towing path, and being ferried across on the barge when the path changes sides. These typical scenes may have gone, but it is still possible to identify the location of many of Constable's paintings, particularly those around Dedham and Flatford Mill.

At Manningtree the river became a wide estuary, bordered by mud flats, mills and small village quays. The river is now permanently separated from the estuary by a tide barrier, but the Stour estuary and connecting Orwell estuary are widely used by boats of all types. Manningtree and Mistley are visited by sailing barges, small coasters and all types of pleasure craft, while Harwich, at the mouth of the river, is traditionally a naval base, and a port for North Sea ferries and freighters. Across the river is Felixstowe, fast becoming the largest container port in England, and rapidly taking the place of Tilbury and the traditional docks of London.

The Orwell and the Gipping
The estuary of the **Orwell** still carries a wide range of boats and ships to Ipswich docks, which give the

town a traditional bustle and variety that is always lacking in those inland port towns and cities that have lost their association with the sea. However, until the 1920s there was another dimension to Ipswich that carried its maritime associations far inland to Stowmarket. The river **Gipping** rises in the rolling Suffolk countryside among the traditional villages of Haughley, Wetherden and Buxhall but it does not achieve any significance until it reaches Stowmarket. The **Ipswich and Stowmarket Navigation** was opened in 1793, allowing a regular traffic to develop along the Gipping through its 15 locks. For well over a 100 years this quiet and undramatic river carried cargoes of agricultural produce, lime, timber, building materials and even explosives between the two towns. Today, the best way to explore the Gipping is on foot as a towpath exists throughout its 17 miles. The remains of many of the locks can be seen, as well as the quays of pleasant towns and villages such as Needham Market. There are also mills and lime kilns, reflections of the navigation's *raison d'être*.

The Suffolk sailing rivers

The Suffolk coastline has three major tidal estuaries that have been associated with boats and sailing for centuries. It is a coastline of mud flats and wild birds, with a strong sense of isolation, and it has witnessed a few major changes in the last 100 years. Sailing for pleasure is the rule instead of sailing for a livelihood, but the essential nature of the area is unaltered.

A detail of a richly-worked purse-lid from the Anglo-Saxon Sutton Hoo Ship Burial. The Sutton Hoo treasures are now in the British Museum.

The **Deben** estuary runs inland through a wooded landscape to Woodbridge, a pretty town built on a low hill where sailing and boats are the major interest. The harbour, which like all Suffolk ports dries out at low tide, is dominated by the restored tide mill. Beyond Woodbridge, the Deben continues through quiet Suffolk villages to its source near Debenham. It was in the Deben marshes near Woodbridge that the Anglo-Saxon Sutton Hoo ship burial was discovered. A few miles along the coast is the mouth of the **Alde**, an eccentric river whose long estuary flows parallel to the coast for some distance. The strip of marshland that separates the river and the sea is a bird sanctuary. Just inland is Orford, whose ruined castle dominates the landscape, and whose oysters dominate the diet of both locals and visitors. At Aldeburgh, formerly a fortified town with a Martello tower still standing on the beach, and now an old fashioned sailing centre, the river turns sharply inland, to pass through woods to Snape, the limit of navigation. By the riverside at Snape is the Maltings, a concert hall and the focal point for the annual Aldeburgh festival established by Benjamin Britten in 1948. Inland the Alde continues to its source near Brundish, passing between the towns of Saxmundham and Framlingham, through a landscape that is the essence of Suffolk.

The third Suffolk river is the short **Blyth**, which rises near Heveningham Hall, one of the finest neoclassical mansions in England. It flows through the traditional market town of Halesworth, meeting the tidal estuary and surrounding mud flats at Blythburgh. From here the estuary can be sailed to its narrow mouth which is flanked by Walberswick and Southwold, the one famous for its associations with the painter Steer, the other renowned, among other things, for its brewery. Although it is hard to believe today, in 1761 the river Blyth Navigation was opened between Halesworth and Southwold and remained in use until the end of the nineteenth century. Many local trading boats used the navigation, aided by its four locks, but in the end the continued silting of Southwold harbour brought about its demise.

The Norfolk rivers and navigations

Although the Norfolk Broads must be generally familiar to everyone, few are aware that the Broads are simply part of a far larger network of over 127 miles of navigable rivers that link many of the major

The entrance hall at Heveningham Hall, one of the loveliest Palladian houses in England. The interior decoration is by James Wyatt.

towns of east Norfolk. This largely natural and self-contained system is without equal in England and for centuries it operated as a vital transport network, linking towns and villages that had no adequate road connections. A special type of sailing barge, the wherry, was developed, capable of reaching the narrowest and most isolated areas. In the nineteenth century hundreds of black-sailed Norfolk wherries could be seen, carrying every type of cargo, but today only one survives in sailing condition. This, the *Albion*, preserved by the Norfolk Wherry Trust, can be seen sailing the Norfolk rivers and Broads during the summer. Remarkably, the network is nearly as extensive as in the nineteenth century, and many areas are still only accessible by water.

Most southerly of the Broads rivers is the **Waveney**, which rises in Thetford Forest where there is a watershed between the Norfolk rivers and those of the Fens. Forming the boundary between Norfolk and Suffolk, the Waveney flows through Bressingham Gardens and Steam Museum, through Diss and a succession of little villages to Bungay, formerly the limit of navigation. Today the limit is Geldeston, whose derelict and isolated lock is guarded by one of the most unusual pubs in England, which as recently as the 1970s had no

Horstead Mill on the river Bure, one of the many attractions along this popular river in the northern section of the Norfolk Broads.

A windy day on the river Thurne, showing a typical Norfolk Broads combination of yacht, cruiser and restored wind pump.

running water or electricity. A pleasantly rural stretch of river leads to Beccles, a largely eighteenth-century town with one of the many large and unusual churches that are a feature of Norfolk, and then to the wide Oulton Broad which connects the river with Lowestoft, and thus with the sea. The Waveney then turns inland, passing the swinging railway bridge at Somerleyton, the ruined priory at St Olaves and the remains of the Roman fort at Burgh Castle, before joining the river Yare at Breydon Water, a huge lake that leads to Yarmouth.

The **Yare** is probably the most important of the Norfolk rivers, for it links Norwich with the sea at Yarmouth. Despite being nearly 30 miles inland, Norwich is still visited by a variety of commercial craft and so the winding course of the Yare is often decorated by coasters and cargo ships, as well as the many types of pleasure boats and sailing cruisers. The combination of a low-lying landscape and a twisting river results in the spectacle of large ships apparently sailing through the middle of fields.

Between Breydon Water and Norwich there are a number of more conventional sights to be seen, including windmills, the twisting course of the river Chet which leads to Lodden, an old market town, the sugar beet plant at Cantley and at Strumpshaw Hall a steam museum. Just outside Norwich the Yare swings away to the south, to continue its unnavigable course to Thetford Forest, and the navigation continues into the city centre on the river **Wensum**. Docks and warehouses, a variety of bridges, the remains of the city wall and the distant view of the cathedral herald the city itself. The navigation ends within sight of a number of Norwich's 32 medieval churches, a suitable place to begin an exploration of one of the finest of England's provincial cities. The Wensum continues its course through the city, and then it meanders in a wide arc through rural Norfolk to its source near East Dereham.

The best known of the Norfolk rivers is probably the **Bure**, which is linked to the rest of the network

Potter Heigham on the river Thurne. Potter Heigham is one of the great boating centres of the Norfolk Broads and its low bridge is a notorious tight squeeze.

by Breydon Water. It has become a pleasure river on a national scale, and many of its villages are devoted to the holiday trade. Boatyards proliferate, and in some areas the banks are lined by summer houses, bungalows and chalets. At the same time the river continues to express the traditional qualities of Norfolk, the reed-lined banks, the succession of Broads linked by river channels, the windmills and windpumps, the attractive and still isolated villages dominated by a rich variety of churches, the birds and wildlife that still survive despite the overcrowding during the holiday season, the waterside pubs and a constantly changing landscape. The most interesting villages along the Bure are Stokesby, Acle, Ranworth, Horning, Wroxham and Coltishall which is the limit of navigation. In the nineteenth century it was possible to continue for a few more miles to Aylsham. The Bure has two tributaries, which expand the navigable network eastwards. The **Thurne** leaves the Bure beneath the sails of a preserved pumping mill, and passes through the holiday and boating village of Potter Heigham to

link some of the most attractive Broads to the network, noted for their wildlife. The limit of navigation is Martham Broad, at which point the sea is less than a mile away. The **Ant** is a narrow and twisting river, flanked by isolated villages, interesting houses, the inevitable boatyards, and the reed beds that still supply many thatchers with their raw materials. Dilham is now the limit of navigation but in 1826 a canal was opened from Dilham to North Walsham. It has been disused since 1935 and is derelict, but the remains of its locks and warehouses can be explored.

The Great Ouse and its connections

Eastern England is blessed by two great river navigations, the **Ouse** and the Nene that not only connect with each other, but with the rest of the English waterways network, and with the North Sea via the Wash. The Great Ouse rises near Brackley in Northamptonshire, and its upper reaches are well worth exploring. An attractive river, twisting its way through the heart of England, it passes Buck-

Ranworth Broad, one of the most remote and least developed part of Norfolk's network of lakes and rivers. It is closed to powered craft.

The church and mill at Olney, Buckinghamshire. Famous for its pancake race, Olney is one of the many attractions along the Great Ouse.

ingham, with the gardens of Stowe nearby; Stony Stratford; the railway town of Wolverton, where it is crossed by the Grand Union Canal in an iron aqueduct; Newport Pagnell and Chicheley Hall, one of the most original early eighteenth-century houses in the country. The river continues past Olney, famous for its pancake race, some fine churches at Felmersham, Pavenham and Clapham and Stevington windmill before it reaches Bedford. The Great Ouse gives added interest to Bedford for here is the head of navigation. From Bedford to Huntingdon the river is pleasantly rural, passing through the market town of St Neots, but from Huntingdon onwards the scenery dramatically improves. A fourteenth-century stone bridge links Huntingdon with Godmanchester, a town of Roman origins, and both have buildings of interest. Houghton Mill is followed by Hemingford Grey,

with its fine waterside church and twelfth-century manor house, while at St Ives the fifteenth-century bridge has a chapel in the centre. At Holywell the Ferry Boat Inn claims to have been established in 980. It also boasts its own ghost, a lady who rises from the floor every year on 17 March. At this point the landscape changes, and from here to the sea the Ouse flows through the flat Fenlands, rich farmlands reclaimed from the sea by the flood prevention schemes of the Dutchman Cornelius Vermuyden in the mid-seventeenth century. Vermuyden was also responsible for creating many of the navigations that can still be enjoyed today. One of these is the Bedford river, a straight canal from Earith to Denver, near Downham Market, that shortens the journey but offers little of interest for the visitor. The old course of the Great Ouse curves round to Ely, via Stretham where a steam pumping engine is preserved, a reminder that the water levels through this part of England have to be strictly controlled to avoid flooding. In some areas the land has sunk 15 feet during the last few centuries, and is still sinking. In this flat landscape, Ely cathedral can be seen for miles. Like Norwich, it is an attractive city when approached by water. From Ely to Kings Lynn the river is broad, fast flowing and of limited appeal, but Kings Lynn itself should not be missed. It is a town of great age and interest, built round a huge market square, and still busy as a port despite the Victorian quality of the docks.

The Great Ouse is also the backbone of a number of other navigations, but those that survive today are only a fragment of the system in its heyday. The **Ivel** is a small river that joins the Great Ouse at Tempsford, just east of Bedford, but in 1823 it was made navigable as far as Shefford, a navigation that survived until the 1870s. Little remains today but the Ivel has a number of aeronautical associations. Near its mouth are the huge airship hangers at Cardington, while during the Second World War the airfield at Tempsford was the base from which

agents were flown into France and other occupied countries on secret missions. A few miles to the south is the Shuttleworth Collection of historical aircraft, at Old Warden. The most important tributary is the **Cam** which links Cambridge, a splendid city to see by water, to the Great Ouse near Ely. North of Ely are three other tributaries but in their present shortened form they offer little of interest. The **Lark Navigation** now comes to a stop in a tiny village, New Row, but until the 1890s it was navigable to Bury St Edmunds. The Navigation had a chequered history which dates back to the 1700s but most of the surviving remains date from the 1890s when an attempt was made to revive it. As late as the 1920s steam tugs could be seen hauling lighters on some stretches. To the north is the **Little Ouse**, or **Brandon** river which was made navigable to Thetford in the late seventeenth century and survived in commercial use until about 1914. Today the limit of navigation is near Brandon, but the upper reaches are worth exploring through Thetford Forest as the river shares a common source with the Waveney and therefore provides a theoret-

The decorative 'Chinese' footbridge at Godmanchester on the Great Ouse, built in 1827. Godmanchester also boasts a fine fourteenth-century stone bridge.

ical link between the Norfolk rivers and Broads and the rest of the English waterway network. Near Kings Lynn the Great Ouse is joined by the **Nar**, a minor waterway rising near Castle Acre Priory. From the mid-eighteenth century to the 1880s this rather unlikely river was also navigable as far as Narborough, and even today the remains of the warehouses and quays can be seen.

The most significant of the waterways linking with the Great Ouse is a series of mostly man-made navigable drainage channels known collectively as the **Middle Level Navigations**. These are remote, low-lying and only visited by the more enthusiastic explorers but they are a hidden part of England that has a quality all its own. More important, they include two routes that link the Great Ouse with the river Nene, and thus with the whole of the English navigable waterway network.

The Nene

The second great river navigation of eastern England is the **Nene**. It rises near Weedon, as far from the sea as it is possible to be in England, and flows through rolling farmland to Northampton, where it becomes navigable. In Northampton the Nene is met by a branch of the Grand Union Canal and is thus linked to the English canal system. The river was first made navigable during the reign of Queen Anne but throughout the eighteenth and nineteenth centuries passage was often erratic and sometimes impossible. In the 1920s the navigation was taken in hand and a whole new series of locks were created, featuring the typical vertically rising guillotine gates. Today it is an enjoyable and easy river to travel along, and is somehow the centre of England as it winds its way among water meadows, and manages to avoid many of the towns and villages en route. It is totally rural, with more cows to be seen than people along the banks. For the same reason, it is a difficult river to explore by car, but local sights include Castle Ashby House, near Cogenhoe, mills at Doddington and Wollaston, near Welling-borough, and the medieval bridge that links Thrapston with Islip. Many of the villages are small and undeveloped, and enjoy fine churches. Near Oundle, a handsome stone town, the Nene begins a series of broad sweeping loops which carry it towards Peterborough. Riverside villages are now

Mepal in Cambridgeshire. Its distinctive waterways and rich variety of wildlife make it a good place to enjoy the landscape of the Fens.

Wisbech, one of the East Anglia's most traditonal inland ports, has a waterfront of rare quality. The seventeenth-century houses reflect a Dutch influence.

more frequent and include Fotheringhay, whose former castle witnessed the execution of Mary Queen of Scots. Near Peterborough is the Nene Valley Steam Railway, which crosses the river twice. Peterborough looks at its best from the river. The old parts of the town, including the cathedral and guildhall are by the river, and the quays are attractive. Leaving the city, the Nene passes the connection with the Middle Level Navigations which lead to the Great Ouse and then the character of the river changes. It becomes a wide waterway flowing through a flat landscape, with a tidal flow that increases as it approaches the Wash. The last town of any significance is Wisbech, an active inland port with one of the most interesting and attractive waterfronts in England. The quays are lined with fine mansions and warehouses, many of which are eighteenth century or earlier. From Wisbech the Nene races along its tidal estuary to the sea, across a landscape that used to be under the sea. Near Sutton Bridge, King John lost the crown jewels.

The Welland

The **Welland** is an isolated river that traverses a region devoted to bulb growing. Navigable now from Deeping St James to its estuary in the Wash, it was in the nineteenth century navigable as far as Stamford via the Stamford Canal which was cut in the 1660s and ran parallel to the river, with 12 locks. Traces of this very early canal can be found at Uffington and at Deeping Gate. South of Stamford is Barnack, the site of quarries which supplied the stone from which many of England's cathedrals and abbeys were built. In the Middle Ages, this stone was transported along the Welland, across the Wash and then to its various destinations via the early river navigations that were the only means of transport until the seventeenth century.

The Witham

The **Witham** and its associated waterways is probably the oldest river navigation in England still in use. It was first developed by the Romans for

Boston Stump on the river Witham. Standing more than 270 feet above the river, it dominates the flat Lincolnshire landscape for miles around.

drainage and navigation, and was then greatly improved during the eighteenth century. The Witham actually rises miles to the south near Melton Mowbray. It flows north, passing Woolsthorpe Manor, the birthplace of Sir Isaac Newton, Grantham House and Belton House, approaches the Trent at Newark and then swings away to Lincoln, where the navigation starts. Like most cities of eastern England, Lincoln makes the most of its river. The Witham passes through a narrow channel flanked by a range of interesting buildings, and the city climbs away to the north, a great slope of stone, brick and timber crowned by the cathedral. East of Lincoln the river is straight and predictable, with high banks hiding much of the landscape but there are features of interest, for example the remains of abbeys near Bardney and Tattershall, Tattershall Castle, and the little town of Woodhall Spa, one of the least known of English spas. In the distance can be seen the so-called Boston Stump, the tower of St Botolph's church which dominates the landscape for miles around. Boston is a fine town and an interesting port, its quays lined with eighteenth- and nineteenth-century buildings. It is a port of considerable antiquity, for the *Mayflower* sailed from here in 1620. There is much to attract the diligent traveller, including a windmill and, nearby, a tractor museum. The Witham estuary beyond Boston links with the Wash, and thus with the other great rivers of eastern England.

Although its best features are the towns along its admittedly unexciting course, the Witham is a river with some interesting connections. A lock in Boston marks the entrance to the **Witham Navigable Drains**, over 50 miles of navigable waterways whose primary function is to maintain the water levels in the surrounding land. This is probably the least known navigable waterway network in Britain, and worth seeing for that reason alone, although the appeal of a flat and isolated landscape, bisected by a series of secret waterways that seem deliberately to avoid all villages, is not universal. More interesting perhaps is the **Fossdyke Canal**, which connects the Witham at Lincoln and with the river Trent near Torksey. This canal, dug by the Romans in about 120 AD, is the oldest man-made waterway still in use, having been improved during the eighteenth century. It provides a vital connection between the Witham and the rest of the English inland waterway network. Until the 1870s there were two other navigations connected with the Witham, the

Lincoln is served by two waterways, the Fossdyke Canal and the river Witham. The Witham has a tortuous passage through the city and under the 'Glory Hole'.

Sleaford and the **Horncastle Canals**. Both opened in 1802, and were planned as part of the same development of the region. Although long closed, their routes can still be traced and the remains of locks discovered. Sleaford and Horncastle are both interesting little-known towns. Traces of the former canal basins can still be found in each, and in Sleaford, the canal company offices survive.

The Louth Canal

One of the least known of the Lincolnshire waterways, the **Louth Canal** was opened in 1770. Like others in the county, its construction was inspired by the constant transport difficulties posed by bad roads and the low-lying marshlands of the region. The canal, with its eight locks, linked Louth on the edge of the Lincolnshire Wolds, with the Humber estuary near Tetney, south of Cleethorpes. In a quiet way the canal prospered and remained in commercial use until the First World War. Although long abandoned, it is still in reasonable condition and its route can be explored. There are locks and warehouses to be seen, and a mill at Alvingham. Louth is an old market town where Lord Tennyson was educated. It has a good church with a tall spire, a rich selection of eighteenth- and early nineteenth-century buildings, and the local museum displays carpets made in the town.

Durham from the air, showing how the city is enclosed by the great sweep of the river Wear. The Cathedral is an outstanding example of Romanesque architecture.

THE NORTH-EAST

The north-east of England is in many ways the most secret part of the country, yet potentially it is the most exciting to explore. The region is formed of contrasting bands, wild and rugged scenery interspersed with thin strips of heavy industry. Each band is full of interest but demands in return a spirit of dedication and adventurousness from its visitors.

The most northerly band is formed by the Scottish borderlands bounded by the Cheviots, the river Tweed and the North Sea, a wild and remote region, filled with reminders of England's early history. Its rivers offer an ideal introduction to the landscape and its past, and a fine way to appreciate the untamed quality of the region. The next band includes the great industrial rivers of the north-east, the Tyne, the Wear and the Tees. These rivers may lack obvious appeal but they are well worth exploring, linking as they do the scenery of the Pennines with the traditional industries upon which England's Victorian wealth and power was based, coal, iron and steel, shipbuilding and international trade. A third, rather broader band comprises the beautiful wilderness of the Yorkshire Dales, and Moors, great tracts of upland divided by the wide valleys of the Derwent, the Ure, the Swale, the Wharfe and the other North Yorkshire rivers. These rivers not only flow through spectacular scenery but they also link together great churches, abbeys, castles and country houses, spanning several centuries of English history. Much of Yorkshire can be explored by water, either by following the river courses or by boat along the rivers that are still navigable. It is still possible to arrive in York by boat several weeks after leaving London.

The valley of the river Coquet in Northumberland, one of the hidden pleasures of England's lesser known rivers.

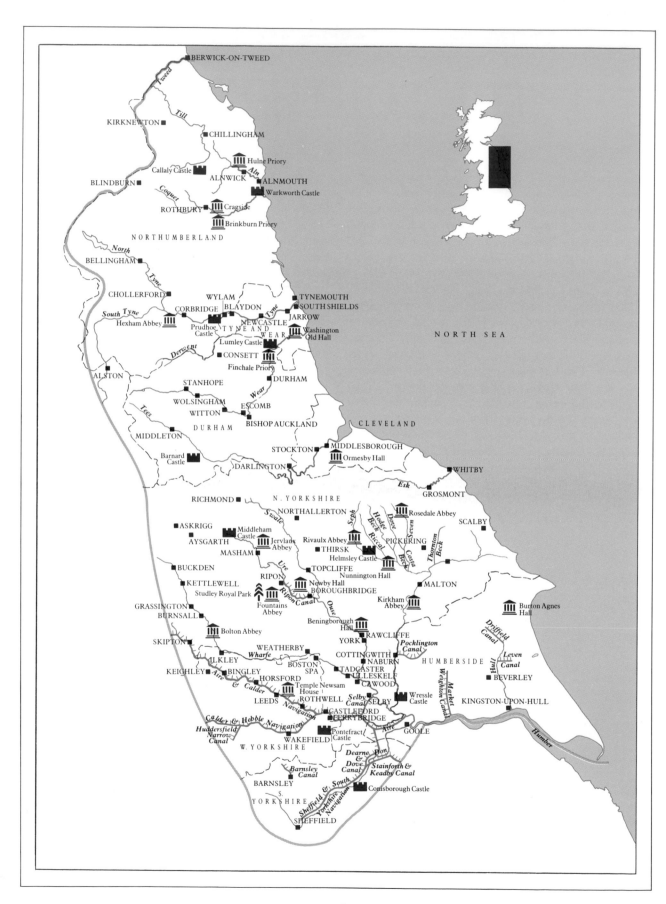

BERWICK-ON-TWEED

KIRKNEWTON

CHILLINGHAM

Tweed

Till

Hulne Priory

Callaly Castle

ALNWICK

ALNMOUTH

Aln

Warkworth Castle

BLINDBURN

Coquet

ROTHBURY

Cragside

Brinkburn Priory

NORTHUMBERLAND

North

BELLINGHAM

Tyne

CHOLLERFORD

WYLAM

TYNEMOUTH

South Tyne

CORBRIDGE

BLAYDON

SOUTH SHIELDS

Hexham Abbey

NEWCASTLE

JARROW

Tyne

Prudhoe Castle

TYNE AND

Washington
Old Hall

Lumley Castle

WEAR

CONSETT

Derwent

Finchale Priory

ALSTON

STANHOPE

DURHAM

Wear

WOLSINGHAM

ESCOMB

WITTON

BISHOP AUCKLAND

CLEVELAND

DURHAM

MIDDLETON

STOCKTON

MIDDLESBOROUGH

Tees

Barnard
Castle

DARLINGTON

Ormesby Hall

WHITBY

Esk

N. YORKSHIRE

GROSMONT

RICHMOND

NORTHALLERTON

Swale

Seph

Rosedale Abbey

SCALBY

ASKRIGG

Hodge Beck

Dove

Seven

AYSGARTH

Middleham
Castle

Jervaulx
Abbey

Rivaulx Abbey

PICKERING

Riccal

Thornton Beck

MASHAM

THIRSK

Helmsley Castle

Costa Beck

BUCKDEN

Ure

TOPCLIFFE

Nunnington Hall

MALTON

KETTLEWELL

RIPON

Newby Hall

Studley Royal Park

BOROUGHBRIDGE

Kirkham
Abbey

Burton Agnes
Hall

GRASSINGTON

Ripon Canal

Fountains
Abbey

Ouse

BURNSALL

Beningborough
Hall

Driffield Canal

SKIPTON

Bolton Abbey

RAWCLIFFE

YORK

Hull

WEATHERBY

Wharfe

COTTINGWITH

Pocklington Canal

ILKLEY

BOSTON
SPA

NABURN

HUMBERSIDE

Leven Canal

KEIGHLEY

BINGLEY

TADCASTER

ULLESKELF

BEVERLEY

Aire & Calder

HORSFORD

CAWOOD

Temple Newsam
House

Market Weighton Canal

ROTHWELL

Selby Canal

SELBY

Wressle
Castle

KINGSTON-UPON-HULL

LEEDS

Navigation

CASTLEFORD

Calder & Hebble Navigation

FERRYBRIDGE

Aire

GOOLE

Huddersfield Narrow Canal

Pontefract
Castle

WAKEFIELD

Humber

W. YORKSHIRE

Dearne & Dove Canal

Don

Stainforth & Keadby Canal

Barnsley Canal

BARNSLEY

Sheffield & South Yorkshire Navigation

Conisborough Castle

S.
YORKSHIRE

SHEFFIELD

NORTH SEA

118

The final band is formed from two quite separate and distinctive regions. The first, Humberside, is a low-lying and little known area of small towns and villages. It is predominantly agricultural, a hidden corner of England overshadowed by the Humber estuary, a great seaway that gives waterway access to the second region, which comprises the industrial towns and cities of Yorkshire. The Humber leads directly to the heart of Yorkshire, an area rich in memorials to nineteenth-century ambition and technical skill. The industrial landscape is impressive and full of interest. Coal and iron are still the dominant forces, their traditional importance underlined by a large network of active commercial waterways. Here, on the Aire and Calder Navigation, on the Sheffield and South Yorkshire canals, at Goole and at Selby, it is possible to relive the atmosphere of the canal age at its height, an experience now unique to the north-east. The canals of Victorian England however, are part of a system of new waterways and docks that are being steadily expanded and improved to enable Continental barges to navigate directly to Sheffield, Doncaster and other centres of industrial activity. This blend of old and new is typical of the north-east and gives it its characteristic flavour.

The rivers of the Border and the Cheviots

The least developed, and correspondingly, the least known part of England is Northumberland. The north of the county is wild, remote and virtually unaffected by the activities of man. The scenery is spectacular and largely undomesticated; great bare hills, tracts of woodland and a mass of small, quick-

The temple of Mithras at Carrawburgh, one of the best preserved Roman buildings associated with Hadrian's Wall.

flowing streams, most of which drain the Cheviot hills. One has to be hardy to survive here and so it is no surprise that the region has produced a number of distinctive, and highly self-sufficient animal species: Cheviot sheep, wild goats, and the Chillingham herd of wild cattle. Among the streams are a number of larger rivers, the most important of which is the **Tweed**. Much of this river is in Scotland, while a long stretch between Coldstream and the estuary actually marks the border. However, Berwick-on-Tweed and the last few miles of the river are in England and a number of interesting tributaries flow southwards into Northumberland. Berwick still has the feeling of a frontier town, underlined by the ruined twelfth-century castle, the town walls and the barracks, which were designed by Vanbrugh in 1717 and are apparently the oldest barracks in Britain still in active use. The **Till** is an attractive tributary of the Tweed, meandering through a valley at the foot of the Cheviots. It is a river surrounded by history. Near its source is Chillingham, which has a fine twelfth-century church and is the home of the white wild cattle, a herd that has lived enclosed in Chillingham Park for over 700 years. The Till passes near Kirknewton with its medieval sculpture, Heatherslaw Mill, and the site of the Battle of Flodden Field, while near its junction with the Tweed is the fifteenth-century Twizel Bridge, the widest medieval single span bridge in Britain. To the south is the river **Aln**, which flows past Callaly Castle, Hulne Priory, and Alnwick with its abbey and Norman fortress on the way to its sandy estuary at Alnmouth.

Perhaps the most dramatic of the Cheviot rivers is the **Coquet**. It rises high above Blindburn, its course through the hills followed closely by a remote

Thomas Bewick's eighteenth-century engraving of the bull of Chillingham, where the herd of wild white cattle have roamed for over 700 years.

road that eventually deteriorates into a track leading to the Scottish border. It is a fast-flowing trout river, like most of those in the region, and makes its way down from the hills into Coquet Dale past isolated little villages. The first town of any size is Rothbury, where the river is dominated by the tall towers of Cragside, a late Victorian mansion designed by Richard Norman Shaw, and the first house in the world to be lit by electricity generated by water power. Nearby, and deep in the wooded river valley, is Brinkburn Priory, which was founded in 1135, ruined in the sixteenth century, restored in the nineteenth century and is still richly endowed with medieval romance. Near the sea the Coquet opens out into a wide estuary, overlooked by the twelfth-century Warkworth Castle which appears to grow out from the precipitous hill. Nearby, cut out of solid rock and standing high above the river is its fourteenth-century hermitage.

The Tyne and its tributaries

One of the major rivers of England, the **Tyne** has been at the centre of English history for the last 200 years. The river witnessed the growth of the Industrial Revolution, the development of railways and the frenetic industrial development of the nineteenth century, when the ironworks, mines and shipyards totally dominated the river banks from Hexham to the sea. More recently the Tyne has watched the decline of these traditional industries as the north-east goes through a period of economic change in every way as far-reaching as the revolution that created the wealth of the region in the eighteenth century. The industrial rivers of the north-east, the Tyne, the Wear and the Tees, are losing their industrial importance but, once they overcome the inevitable tide of decay and dereliction, they may begin to return towards their pre-industrial state. Certainly all three are already rivers

Canoeing on the river Wear. The violent and dramatic courses of many northern rivers makes them particularly challenging for canoists.

of extraordinary contrast, flowing from their sources high in the Pennines through dramatic scenery and wild beauty, with no hint of the total dedication to industry that determines their nature a few miles further towards the east coast.

The Tyne enjoys two sources, which come together to form one large river near Hexham. The **North Tyne** rises in Kielder Forest, flowing from the huge Kielder Water reservoir, which is used extensively for water sports, leisure activities and fishing. It is a fairly remote river, passing through Bellingham and a number of smaller villages where it meets tributaries that drain the Wark Forest to the west. At Chollerford it is crossed by Hadrian's Wall, which is in a particularly impressive state in this area. Nearby are forts, notably Brunton Turret, a bath house at Chesters and a temple at Carrawburgh, sturdy reminders of the level of civilisation attained by the Romans in such primitive and inhospitable regions. The **South Tyne** follows a very different course, rising in the Pennines far to the south, by Alston Moor. Despite the remoteness of the region, exploration is not difficult. The river is accompanied by two little roads and by the Pennine Way and until fairly recently there was also a railway to Alston. Alston is an attractive hill village and market town, which is frequently cut off during the winter. There are a number of smaller villages beside the river on the way to Haltwhistle, where it becomes more accessible. There are plenty of castles, Blenkinsop, Bellister and Langley, reflecting the turbulent history of the region and just west of Hexham the South Tyne joins its northern

The transporter bridge across the Tees at Middlesborough is a unique survival of a once common example of Victorian engineering.

branch to become the Tyne. At Hexham the industry begins, but there are still a number of reminders of the pre-industrial Tyne, Hexham Abbey, the Saxon church at Corbridge and the fourteenth-century Prudhoe Castle. Industrial history is also well represented, there is the National Tractor Museum at Hunday and Wylam is the birthplace of George Stephenson. This mixture of medieval and industrial history continues along the banks of the Tyne as it widens into a tidal estuary, passing Newcastle, Jarrow and South Shields before meeting the sea at Tynemouth. A similar type of river is the **Derwent**, the major tributary of the Tyne. This rises in the wooded Derwent Reservoir and enjoys a short scenic route before meeting industry at Consett. Beyond Consett the Derwent takes an attractive, wooded course through Rowland's Gill, passing the ruined Gibside Estate, with its Elizabethan Hall, orangery, gardens and its eighteenth-century chapel with its fine three-decker pulpit. The Derwent joins the Tyne near Blaydon, at one time the home of the famous races.

The Wear

The **Wear** is in many ways a similar river to the Tyne. It also rises in the Pennines, to the south-east of Alston, in an area that was formerly the scene of extensive mining. Only old shafts and quarries remain, now looking a natural part of the landscape. Between St John's Chapel, Stanhope and Wolsingham the river passes through the scenic Weardale, where the fast-flowing waters are popular with canoeists. Beyond Wolsingham the Wear starts a huge loop that takes it southwards through Witton

The upper reaches of the Tees are characterised by waterfalls. The most spectacular of these is High Force, near Middleton-in-Teesdale.

Whitby harbour is attractively old-fashioned, its quays and fishing boats reflect the traditional flavour of the English seaside.

with its castle, Bishop Auckland and Escomb Saxon church before turning north towards Durham. The view of the Wear curving round Durham, with the cathedral raised high above on a steeply wooded hill is justly famous. It is relatively unchanged since the eighteenth century and provides an exciting introduction to a fine city. Despite the industrial nature of the region as a whole, the Wear manages to retain its historical associations and it has a predominantly rural nature. North of Durham much of its course is wooded and there are the remains of Finchale Priory and Lumley Castle, the latter, like so many of the country houses in this area, provides the setting for Elizabethan banquets and similar entertainments. Washington Old Hall, a Jacobean mansion that has earlier connections with the Washington family, links the Wear to the United States, and nearby is a wildfowl reserve, marking the start of the Wear's

estuary. From here to the sea, industry and particularly ship-building have determined the nature of the Wear, although many other industries used to be associated with Sunderland, including the making of pottery and glass.

The Tees

The third of the great rivers of the north-east is the **Tees**. Rising in the Pennines, in the remote wilderness of Moor House Bird Reserve, the Tees flows into Cow Green Reservoir, which it leaves via the dramatic Caldron Snout waterfall and cataract. At the base of the waterfall its course is joined by the Pennine Way, which accompanies it for several miles through Teesdale to Middleton. There are several waterfalls of which the most spectacular is the High Force, the highest in England, and they can all be seen from the Alston-Middleton road, which follows the route of the Tees. After Middleton, the river becomes less dramatic, its wooded course leading to Barnard Castle. A fine town built high on a hill overlooking the river, Barnard Castle has a number of attractions, including the castle itself, Egglestone Abbey and the Bowes Museum, a rich and varid collection of works of art housed in a huge Victorian version of a French château. Between Barnard Castle and Darlington the Tees follows a rural course, forming the border between Durham and Yorkshire. East of Darlington the river becomes navigable, starting a 24 mile tidal estuary through Stockton and Middlesborough to the sea in Hartlepool Bay. This section is busy with shipping, and the banks are heavily developed, with steel, chemicals, ship-building and other industries. Despite these rather daunting surroundings, there is still plenty of interest, notably eighteenth-century Ormesby Hall, the Captain Cook Museum, and the Middlesborough transporter bridge, a unique example of Victorian engineering.

The Esk

One of the lesser known rivers of the north-east is the Esk, a salmon river that rises in the Cleveland Hills and flows eastwards in a steep valley through the North Yorkshire Moors. It can be explored by road but the best way to enjoy it is by rail. The branch line from Middlesborough to Whitby shares its valley and is a particularly scenic journey, while at Grosmont, one of a number of small riverside villages, there is a junction with the North Yorkshire Moors Steam Railway. The Esk meets the sea at Whitby, an attractive town with an abbey, a museum, a beach and a good old-fashioned atmosphere, particularly around the harbour.

The rivers and canals of Yorkshire

The river **Humber**, a great tidal waterway that carves its way deep into the north-east of England, is not only one of the great shipping rivers of England but it is also the backbone of a great network of rivers and canals that radiate throughout Yorkshire. Wherever they rise, most Yorkshire rivers make their way down to the Humber estuary by one means or another. The Derwent, the Ouse, the Ure, the Wharfe, the Swale, the Hull, the Aire and the Calder all mingle their waters in the Humber. For many centuries the very size of the Humber made it a natural barrier, dividing Yorkshire from Lincolnshire and the north Midlands and marking the

distinctive change in the landscape. It is a huge river, wide and fast flowing, its banks isolated and relatively undeveloped. Until the opening of the Humber suspension bridge which incorporates the world's largest span (4626 feet between piers) it could only be crossed by ferry. Even today, wildfowl and bird reserves are as much a part of the Humber as oil terminals, container ports and shipyards. The importance of the Humber as a shipping river is also reflected by the many canals and waterways that connect with it. A great number of Yorkshire's rivers were turned into navigations and many are still open and active today, while the remains of others long closed or abandoned reveal how extensive the network was in its heyday. From the Humber boats can travel northwards to York

The Humber suspension bridge. It is the most recent and the most dramatic of the many exciting river crossings to be found in England.

and beyond, westwards to Leeds and thence across the Pennines to the narrow canals of Manchester and the west of England, and southwards along the Trent to the Midlands and ultimately to London.

Because they are so numerous and so varied, the rivers and canals of Yorkshire cannot be described as one great network, but must for convenience be separated into smaller groups or units.

The Derwent and its tributaries

The **Derwent** rises in the high moors west of Scalby and flows towards the sea before turning inland along the wide glacial valley that leads to Malton. North of Malton the Derwent is joined by a network of tributaries, all flowing south from the North Yorkshire Moors. This collection of fast-flowing rivers and streams leads not only to spectacular countryside and quiet, hidden valleys, but also to some remarkable buildings. The **Thornton Beck** leads to Stain Dale and Dalby Forest; the **Costa Beck** to Pickering and its castle; the **Seven** to Rosedale Abbey; the **Dove** to Farndale, the Dale of the Daffodils; the **Hodge Beck** to Bransdale, a land-locked valley; the **Riccal** to Helmsley Castle and seventeenth-century Nunnington Hall; and the **Seph** to Rievaulx Abbey, one of the most beautiful ruins in England. Exploration of these moorland rivers reveals many other delights, which are accessible by minor roads along the river valleys. Malton, a town with Roman associations, marks the former head of navigation of the river Derwent.

Opened in 1701, and briefly extended for a further 11 miles to Yedingham, the navigation ran for 38 miles to the junction with the Ouse at Barmby-on-the-Marsh. There are five locks, but the upper section from Stamford Bridge to Malton was closed in 1935. The Derwent is a particularly attractive waterway and so it is likely that the navigation will be restored through to Malton. It is a quiet river, flanked by small villages but there are plenty of historical associations, for example the ruins of Kirkham Abbey and Wressle Castle, and the site of the Battle of Stamford Bridge, won by King Harold before he marched southwards to meet his doom at the Battle of Hastings.

The Pocklington Canal

The **Pocklington Canal**, a pleasantly obscure waterway with nine locks, leaves the river Derwent at Cottingwith to wander gently towards Pocklington. Opened in 1818 it continued in use until the 1930s and then fell into decay. Although its route is through a remote landscape of marshland and fens,

The dramatic ruins of Rievaulx Abbey by the river Seph. The chief glory of the abbey now is the thirteenth-century choir, a splendid example of northern English architecture.

A pretty bridge over the remote Pocklington Canal in Yorkshire, a nine-mile waterway gradually being restored by volunteers.

it is an attractive canal, with well-built bridges and locks, its surroundings rich in bird life. The route of the canal is easy to follow and restoration plans make it likely that boats will one day be able to reach Pocklington again.

The Ouse and its connections

The **Ouse** is the major waterway of Yorkshire. A river of great contrasts, its waters flow from Wensleydale through constantly changing scenery to its junction with the Humber and the Trent at Trent Falls. The Ouse is actually two rivers for, from Ouse Gill Beck, a few miles north of York, it changes its name to the **Ure**, although the two are effectively the same river. The Ure rises near Askrigg in Wensleydale and takes a dramatic course through the Dales, tumbling over waterfalls, notably at Aysgarth, and passing the ruins of Middleham Castle and Jervaulx Abbey before reaching Masham. From here the river pursues a more sedate and rural route to Ripon, a fine and little known cathedral town, with Fountains Abbey and the Studley Royal landscape park and gardens nearby.

Ripon is linked to the Ure by the **Ripon Canal**, opened in 1773 to bypass a shallow and rocky stretch of the river. There are three locks, the upper two of which were abandoned in 1955, leaving the present terminus near Ripon race course. However, the canal can easily be followed to its original basin near the cathedral. Ripon is the most northerly point on the English inland waterway network; a boat setting off from Godalming on the river Wey, south of London, could moor at Ripon several weeks later. South of Ripon the Ure continues through beautiful country, passing Newby Hall and flowing through Boroughbridge, which once boasted 22 coaching inns, to its junction with the Ouse at Swale Nab. The Ouse winds its way through fields and woodlands, passing the early eighteenth-century Beningbrough Hall on its way to York. York is still an active inland port and its traditional waterfront is particularly attractive, with a variety of early warehouses and mills, and an interesting selection of commercial and pleasure boats. South of York the Ouse continues through rural surroundings to Naburn, where its character begins to change. From Naburn

The river Ouse at York. The city built its prosperity on the river trade and its waterfront is a rich and varied architectural delight.

southwards the river is tidal and it becomes increasingly wild and unfriendly as it approaches the Humber. The muddy banks give way to bleak marshland that the river flows through at an alarming speed, apparently reaching nine knots in places. The river demands great care in navigation but there is plenty to be seen from the land. Cawood is a pretty riverside town, while Selby and Goole are busy inland ports, which retain the atmosphere of the Victorian period when their importance was first established. Selby, with its fine market place and twelfth-century abbey, is still predominantly a waterway town. The Selby Canal links the Ouse to the **Aire and Calder Navigation**, one of the busiest commercial waterways of the north-east. Goole also has a connection with the Aire and Calder, and indeed Goole was established by the Aire and Calder Company as a major canal port in the nineteenth century. South of Goole the Ouse becomes more like the sea than a river, with sandbanks, lighthouses and a wild tidal flow, rushing through a dramatic wilderness surrounded by marshland.

A number of other rivers of interest connect with the Ouse, notably the **Swale** and the **Wharfe**. The Swale rises, like the Ure, in the Dales and its upper reaches as far as Richmond are similar in character. Richmond is one of the most delightful small towns in England and it makes the most of its river, but south of Richmond the Swale becomes a curiously remote rural river, its banks unapproached by villages or roads. Despite its isolation, there were plans in the middle of the eighteenth century to turn the Swale and some of its tributaries into navigations, and thus link Northallerton and Thirsk to the Ouse. Some work was carried out during the 1760s, making the Swale navigable to Topcliffe and constructing a basin and wharf at Thirsk, but it is not likely that these proposed navigations were ever completed. The **Wharfe** is altogether a more attractive river, rising in Langstrothdale high in the Dales. It follows an exciting course through Wharfedale and offers many features attractive to canoeists. Its route through Wharfedale is followed by a road and the Dales Way footpath, making exploration easy and enjoyable. There are many pretty riverside villages, notably Buckden, Kettlewell, Grassington and Burnsall, there are the ruins of Bolton Abbey and, as the Dales descend to the moors, there is Ilkley. East of Ilkley, the river enters a more cultivated landscape as it skirts to the north of Leeds, passing through Weatherby and Boston

Spa on its way to Tadcaster. From Tadcaster to its junction with the Ouse at Cawood the Wharfe has periodically been navigable, although no formal navigation works have ever been carried out. Certainly trading barges regularly reached Tadcaster during the eighteenth century, and as late as the 1890s some Tadcaster brewers formed a company to improve the navigation. Today the river is often shallow and impassable above Ulleskelf, the tidal limit.

The canals and rivers of Humberside

A number of waterways flow into the north bank of the Humber, including some early navigations that are now mostly disused. One of the more remote of these is the **Market Weighton Canal** which was opened in the 1780s, partly to link Market Weighton to the Humber, and partly to drain the flat farmlands below the Yorkshire Wolds. Despite its undramatic quality this canal remained in commercial use until early this century, when the top three miles were closed, although the lower section was still in use as late as the 1950s. Since then, the canal has had a somewhat chequered history. In 1971 the sea lock

into the Humber was closed, thus closing the canal, but in 1978 the lock was restored and back in use, thus reopening the lower six miles of the canal. The route of the upper section to Market Weighton can still be traced, and a number of original canal buildings survive.

A more significant waterway is the river **Hull** which rises in the Wolds near Burton Agnes Hall and then flows south through Beverley to join the Humber at Kingston upon Hull. Twenty miles of this river, effectively the tidal section, are still navigable, and so the pleasant town of Beverley can be visited by boat. The landscape is not very exciting but the river has a number of connections with Hull docks and so there is plenty of activity. Two small canals link with the river Hull, the **Driffield Canal** and the **Leven Canal**. Of these, the former is the more interesting. Opened in 1770 and active until the 1940s, the canal has a number of attractive locks, some swing bridges and, at Driffield itself, a very attractive basin surrounded by fine early warehouses. Easy to explore, and extensively used by small boats, the Driffield Canal is well worth a visit.

An aerial view of Goole, showing the complex network of rivers, canals and docks that underline the town's importance in the Victorian era.

The commercial waterways of Yorkshire

Yorkshire still boasts a network of busy, commercial waterways making it, in effect, the only place in England where a great variety of trading vessels can still be seen. Here is the only English parallel with the commercial waterways of Europe, for it is only in this area that the waterways are sufficiently large and well-maintained to attract commercial traffic both from the Continent and from other parts of England. This network links Leeds, Wakefield, Castleford, Doncaster, Sheffield and other industrial centres with Goole, and thus with the Humber and the North Sea. The basis of the network is three rivers which have been navigable in part for several centuries: the **Aire**, the **Calder** and the **Hebble**. During the nineteenth century these navigations were greatly improved and expanded and the waterways of Yorkshire were among the few in England able to overcome the threat posed by the railways. Through the nineteenth century and into this century commerical traffic continued to increase, and some of these waterways still hold their own against competition from road transport.

The Aire is the largest of the three rivers, and also the most attractive. Rising in the Dales north of Skipton, it is for the first few miles a typical uplands river, fast flowing and scenic. It follows closely the course of the Leeds and Liverpool Canal, through Airedale to Keighley and then to Bingley, Horsford and Leeds itself where it becomes a navigation in its own right. From Leeds the Aire continues through Rothwell, Castleford, Ferrybridge and Rawcliffe to its junction with the Ouse north of Goole. Although still navigable, the winding course of the Aire between Ferrybridge and the Ouse has been bypassed by an artificial canal which takes a direct route to Goole. A branch canal leads to Selby and another, following the course of the river Calder to Wakefield, connects with the Calder and Hebble Navigation and continues to Sowerby Bridge, where it ends. Originally two trans-Pennine canals, the Rochdale and the Huddersfield Narrow, connected with the Calder and Hebble, thus linking Manchester with Leeds and the north-east, but these have been closed for many years (for further details see *The North-West*).

A view of the river Swale near Keld, showing North Yorkshire Swaledale scenery at its best.
The Swale is an attractively remote rural river.

A train of 'Ton Pudding' compartment boats. They have been used for transporting coal on the Aire and Calder Navigation since the Victorian period.

Although these waterways are predominantly industrial, there is plenty to see. The landscape itself is often quite dramatic, blending well with the powerful relics of Victorian development. There are also many echoes of an earlier period of civilisation, for example Wakefield Cathedral, Pontefract Castle and eighteenth-century Temple Newsam House. However, for many people the greatest interest is the commercial traffic that is still using these waterways. Barges loaded with chemicals, oil and petrol, industrial products, gravel and stone and above all coal can still be seen in plenty, particularly on the Aire and Calder Navigation. Coal was always the major traffic and even today a number of power stations are still supplied by water. The best place to watch this traffic from is Ferrybridge in West Yorkshire. Here the Tom Puddings can be seen, trains of compartment boats full of coal, each of which is physically lifted out of the water to be emptied at the power station by a huge mechanical hoist providing another example of Victorian ingenuity that is still in action.

The other great commercial waterway of Yorkshire is the river **Don**. The Don Navigation was opened from 1751 to 1819 to link Sheffield and Rotherham with Goole and the Humber estuary. Like the Aire and Calder, this navigation was continuously expanded and improved and so was

able to remain successful throughout the nineteenth century and well into the twentieth century. In 1895 it was amalgamated to form the **Sheffield and South Yorkshire Navigation**, with branches connecting with the Aire and Calder, with the river Trent via the Stainforth and Keadby Canal and with Barnsley and Wakefield via the Deane and Dove and Barnsley Canals. The **Dean and Dove** and **Barnsley Canals** were closed during the 1950s and 1960s, and although parts of their route have been obliterated, they can still be traced. However, the other parts of the network continue to thrive, and the section between Doncaster and Rotherham has recently been entirely rebuilt and enlarged to take continental-sized barges and to enable commercial traffic to sail directly between the Continent, the Humber ports and the industrial centres of Yorkshire. This rebuilding programme, completed in 1982, represents the first major investment in commercial waterways in England since the 1930s. The Sheffield and South Yorkshire is above all a commercial network carrying heavy traffic through a predominantly industrial environment but it is not without interest. It has plenty of historical associations and a number of surprises. Among these are Conisborough Castle, a splendid twelfth century castle whose massive keep still looks nearly new, and the trolleybus museum at Sandtoft.

Liverpool's waterfront with the Liver Building, topped by its famous birds, the most traditional and familiar view of the Mersey.

THE NORTH-WEST

The north of England is divided into two by the Pennines, a great chain of hills which forms the geographical and physical backbone of the country. The Pennines are also a major watershed, separating the rivers that flow eastwards to the North Sea from those that fall westwards to the Atlantic. The north-west of England is a distinctive and highly varied region, its nature determined to a considerable extent by the waterways that run through it. The greatest of these waterways are the Eden, the Lune and the Ribble. These rivers are little known but their long courses warrant exploration for they reflect the changing character of the north-west itself, a landscape that ranges from the dramatic and isolated scenery of the north, full of echoes of Roman and medieval England, to the industrial heartlands of the south.

In the far north, the Scottish borderlands that surround Carlisle are not well known, yet they contain a number of pleasant rivers that follow quiet routes to the Solway Firth. Far better known are the waterways of the Lake District and the Cumbrian mountains, which have been attracting visitors since the eighteenth century. The pleasures of the Cumbrian coast should not be overlooked and they are easily explored via its rivers. To the south lies more splendid scenery, notably the Yorkshire Dales and the Forest of Bowland, and the contrasting coastline around Morecambe Bay, associated for centuries with fishing and shipbuilding. The waterways of south Lancashire reflect the impact of the Industrial Revolution. This area was radically changed by the growth of the textile industry, which was dependent upon rivers and canals for power and transport. Centres of industrial activity grew up along these waterways, creating a pattern still apparent today despite the changed nature of the industry and its environment. The Lancashire towns on the foothills of the Pennines are full of echoes from the eighteenth and nineteenth centuries, and the best way of

Cows grazing at Camboglanna Roman fort at Birdoswald, one of the best preserved Roman buildings associated with Hadrian's Wall.

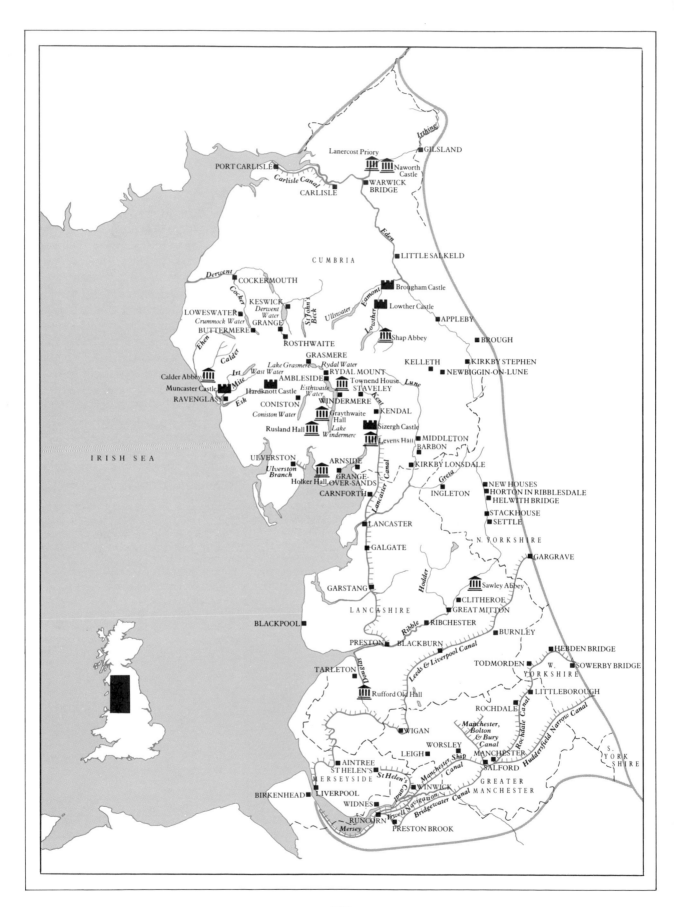

PORT CARLISLE
Carlisle Canal
CARLISLE
WARWICK
BRIDGE
Lanercost Priory
Naworth Castle
GILSLAND
Irthing

Eden

CUMBRIA
LITTLE SALKELD

Derwent
COCKERMOUTH
KESWICK
Cocker
Derwent Water
LOWESWATER
Crummock Water
GRANGE
BUTTERMERE
ROSTHWAITE
St John's Beck
Ullswater
Eamont
Lowther
Brougham Castle
Lowther Castle
APPLEBY
Shap Abbey
BROUGH
Ehen
Calder
GRASMERE
Lake Grasmere
Rydal Water
RYDAL MOUNT
KELLETH
KIRKBY STEPHEN
NEWBIGGIN-ON-LUNE
Irt
Wast Water
Calder Abbey
AMBLESIDE
Townend House
STAVELEY
Lune
Muncaster Castle
Mite
Hardknott Castle
Esthwaite Water
WINDERMERE
RAVENGLASS
Esk
CONISTON
Coniston Water
Kent
KENDAL
Graythwaite Hall
Rusland Hall
Lake Windermere
Sizergh Castle
MIDDLETON
BARBON
Levens Hall
KIRKBY LONSDALE
ULVERSTON
Ulverston Branch
ARNSIDE
GRANGE-OVER-SANDS
Holker Hall
Greta
NEW HOUSES
HORTON IN RIBBLESDALE
HELWITH BRIDGE
CARNFORTH
INGLETON
STACKHOUSE
SETTLE

I R I S H   S E A

Lancaster Canal
LANCASTER
N. YORKSHIRE
GALGATE
GARGRAVE
Hodder
GARSTANG
Sawley Abbey
CLITHEROE
GREAT MITTON
L A N C A S H I R E
RIBCHESTER
Ribble
BLACKPOOL
BURNLEY
PRESTON
BLACKBURN
HEBDEN BRIDGE
Leeds & Liverpool Canal
TODMORDEN
SOWERBY BRIDGE
W. YORKSHIRE
TARLETON
Douglas
Rufford Old Hall
LITTLEBOROUGH
ROCHDALE
Rochdale Canal
Huddersfield Narrow Canal
WIGAN
Manchester, Bolton & Bury Canal
S. YORKSHIRE
LEIGH
WORSLEY
MANCHESTER
AINTREE
ST HELEN'S
Manchester Ship Canal
SALFORD
GREATER MANCHESTER
St Helen's Canal
MERSEYSIDE
WINWICK
BIRKENHEAD
LIVERPOOL
Irwell Navigation
WIDNES
Bridgewater Canal
RUNCORN
Mersey
PRESTON BROOK

appreciating this is to travel along the waterways, especially the great trans-Pennine canals.

The wide estuary of the Mersey forms a sheltered seaway that enabled Liverpool to develop from the seventeenth century as a centre for international trade, particularly with America and the colonies, and ships are still a vital part of Liverpool's economy. Liverpool's neighbour, Manchester, is a great Victorian metropolis whose wealth was drawn from the corners of the British Empire. This corner of England is currently experiencing a painful period of change as the traditional pattern of industry is slowly dismantled, and now is the time to explore it, before centuries of history disappear. The waterways of the region are a direct historical link. The Duke of Bridgewater's canal was the first modern canal to be built in England and it inspired a network of canals that are now largely forgotten. There is still much to be discovered in the network of narrow canals that linked Manchester to the rest of England, in the docks, the Mersey, and along the Manchester Ship Canal, which connected British industry to the rest of the world.

The Eden and its tributaries

Rising high in the Pennines east of Brough, the **Eden** is for many miles a wild and remote river. For much of its course it drains the Pennines, flowing quickly through a dramatic landscape of hills, moorland and woods. Despite its relative isolation, the Eden is not a difficult river to explore. Minor roads frequently run close to it, but the best way to see it is from the Settle to Carlisle railway line. This is one of the most exciting railway journeys in Britain but the line is now threatened with closure. The wildness of the river is matched not only by its landscape, but also by the towns and villages that flank it. There are castles at Brough, Appleby and Corby, all built originally as medieval fortresses, while a stone circle near Little Salkeld is a reminder of a far earlier period of civilisation. Although there are a few old mills along its course, the Eden is a very undeveloped river and salmon and trout can still be seen among its rapids. The only town of any significance is Carlisle, traditionally a fortified town and still a famous garrison. Although rather gaunt, Carlisle wears its past well. There is a cathedral, a castle, a museum and plenty of strong stone buildings. West of Carlisle the Eden follows a more rural course, undulating gently to its tidal estuary on the Solway Firth.

One of the best known known features of northern England and the Borders are the Roman remains, and in particular Hadrian's Wall. A good way to explore the Wall and its associated musuems is to follow the river **Irthing**, for it is never far from the Wall on its way to join the Eden near Warwick Bridge, east of Carlisle. The Irthing rises in the

Lanercost Priory, one of the many historical landmarks along the course of the river Irthing. The Priory was founded in 1169.

inaccessible Wark Forest and then joins the Wall near Gilsland, passing two forts, the most famous of which is Birdoswald. However, not everything is Roman, for nearby is Naworth Castle and the romantic ruins of the twelfth-century Lanercost Priory, set in woods on the north bank of the river. Another interesting tributary is the river **Eamont** which connects Ullswater to the river Eden. It also has historical associations, notably the prehistoric earthwork known as King Arthur's Round Table, near Eamont Bridge, and nearby the ruins of Brougham Castle. Eamont Bridge marks the junction of the Eamont and the **Lowther**, the latter leading to the huge Haweswater reservoir, the isolated ruins of Shap Abbey and Lowther Castle with its wildlife park.

The Carlisle Canal

One of the most forgotten features of Carlisle is the **Carlisle Canal**, which once linked the town of Port Carlisle on the Solway Firth. This ambitious attempt to develop Carlisle into a seaport dates from the early 1820s. A wide canal with 8 locks, it enjoyed a modest success in transporting coal and building materials for 30 years. There was also a passenger service along the canal and travellers bound for Liverpool could spend the night in a specially built hotel in Port Carlisle. In 1853 the canal was closed and replaced by a railway, which in turn has disappeared. Traces of the canal can still be found, including warehouses and a fine customs house in Carlisle, and a lock, basins and wharves in Port Carlisle, as well as the hotel building formerly used by Liverpool-bound passengers.

The Derwent and the Northern Lakes

The Lake District is one of the most familiar leisure areas in England, its many lakes, rivers and streams have been enjoyed by generations of visitors. It is a region of mountains and high hills, with many large expanses of enclosed water, linked and drained by innumerable rivers and streams. The rivers tumble rapidly through steep, rocky valleys and there are plenty of dramatic waterfalls, or forces. Much of it is inaccessible except to the most dedicated and well-equipped of explorers, but there is still plenty to see and enjoy along the main rivers which connect the lakes together.

The major river of the region is the **Derwent**, which rises high in the mountains above Borrowdale. Tiny wooded villages, notably Rosthwaite and Grange, mark its course to Derwent Water, after which it flows on through remote woodlands to the long Bassenthwaite Lake. To the west is Cockermouth, a little town famous for its church and the house where Wordsworth was born in 1770. West of Cockermouth the Derwent becomes a placid river, wandering through farmland to its mouth near Workington, a traditional mining town with a

Autumn colours near Keswick, one of the most popular Lake District resorts.

strong Victorian flavour. The other northern lakes are connected to the Derwent by its tributaries. **St John's Beck** and the river **Greta** lead to the thickly wooded Thirlmere, a long narrow lake at the foot of Helvellyn mountain. The Greta also passes through Keswick, a popular Lake District resort whose local museum has a collection of Wordsworth mementos. Cockermouth marks the junction between the Derwent and the **Cocker**, the latter flowing through Lorton Vale to link Loweswater, Crummock Water and Buttermere. West of Buttermere village is the dramatic Scale Force, a waterfall that plunges over 120 feet through the rocks. To the south-east lies Seathwaite, reputedly the wettest place in England, with an average rainfall of over 160 inches.

The rivers of the Cumbrian coast

The cliffs and sand dunes of the Cumbrian coast between Whitehaven and Barrow-in-Furness are intersected by a number of small rivers which flow busily down to the sea from the mountains and fells.

A typical Lakeland scene, the valley of St John's Beck between Keswick and Thirlmere. This area abounds in superb walks.

Hire boats are available on the larger lakes and provide a good way to enjoy the scenery.

The **Ehen** links Ennerdale Water with Cleator Moor and Egremont, while the **Calder**, which shares its estuary with the Ehen, passes the ruins of Calder Abbey. The **Irt** flows west from Wast Water, the deepest lake in England, overlooked by Sca Fell. Between the rivers **Mite** and **Esk**, which flow into the sea either side of Ravenglass, is the track of the narrow gauge Ravenglass and Eskdale steam railway, its climbing route built originally in 1875 to transport iron ore to the coast. The Esk also flows past Muncaster Castle, a fortified house dating back to the thirteenth century, famed for its furnishings, ornamental gardens and tropical birds. Nearer its source, and 800 feet above the Esk is a much older structure, the Roman Hardknott Castle, built to defend England against the Picts and the Scots prior to the construction of Hadrian's Wall.

The southern Lakes

Flowing into Morecambe Bay from the north are the rivers that lead directly to the best known group of lakes in England: Coniston Water, Windermere, Esthwaite Water, Rydal Water and Grasmere. Apart from the natural pleasures of the landscape, this region offers a wide variety of other interests. The pretty lakeside towns and villages of Coniston, Windermere, Ambleside and Grasmere have managed to preserve their identity despite the thousands of visitors. There are fine houses and gardens, notably Townend House, Graythwaite Hall, Brockhole, Court House and Rusland Hall, and interesting literary associations, such as the Wordsworth Museum at Grasmere and his house at Rydal Mount, and John Ruskin's house at Brantwood, overlooking Coniston Water. For those attracted by the history of transport there is a steam railway, the Lakeside and Haverthwaite, at the southern end of Windermere, and Windermere also has a Steam Boat Museum. In this region of lakes there are unusually few rivers of significance, but the lakes themselves are easy to explore and enjoy. The best way to see them is by boat. On some, such as

Arnside on the estuary of the river Kent in Cumbria. The Kent estuary is flanked by wide sandy beaches, and Arnside is a popular holiday resort.

Originally built to carry iron ore, the miniature Ravenglass and Eskdale steam railway is now a major tourist attraction on the Cumbrian coast.

Windermere and Ullswater, there are scheduled passenger services, and on many others small boats, dinghies, canoes and sailing craft, can be hired.

To the south of the Lake District lie Cartmel Sands, which are overlooked to the east by Holker Hall with its gardens, motor museum and adventure park, and to the west by the town of Ulverston. For many years Ulverston was linked to Morecambe Bay by a short ship canal. Opened in 1796, this gave the town a new life as an inland port and it remained active until early this century. The **Ulverston Canal** was finally closed in 1945, but still survives virtually intact, and so can easily be explored.

The Kent
Although lacking the obvious drama of many of the Lake District rivers, the **Kent** should not be overlooked. It is relatively accessible throughout its course and passes a number of features of interest. Rising above the Kentmere Reservoir high in the mountains, the Kent is for its first few miles a typical uplands stream, wild and fast flowing. It passes

through Staveley, where traditional country furniture is made and then flows through a more open, agricultural landscape to Kendal. This town is the gateway to the Lake District and its many attractions include a castle, the eighteenth-century Abbot Hall Art Gallery and Museum, with its collections of furniture and paintings, the Lakeland Life and Industry Museum, and the local museum of natural history and archaeology. The Kent flows through the town, adding to its appeal. South of Kendal the river passes along a wide valley on its way to its estuary into Morcambe Bay, flanked at its mouth by Arnside and Grange-over-Sands. Although busy main roads run close to the river in this area it is still worth exploring. Two major country houses dominate the valley: Sizergh Castle, an Elizabethan mansion with a fourteenth-century tower and a seventeenth-century garden, which has been the home of the Strickland family for over 700 years; and Levens Hall, famed for its topiary garden, its furniture and, more unexpectedly, its collection of steam traction engines.

Bridge House, Ambleside, one of the most distinctive architectural features in this attractive village that has remained largely unspoiled despite its many visitors.

The Lune

The river **Lune** rises high in the Fells west of Kirkby Stephen and effectively marks the transition between the dramatic landscape of the Lake District and the more domestic countryside of north Lancashire. At first wild and remote, the Lune soon enters the wide valley which it shares with the M6 motorway and the main west coast railway to Scotland. This valley leads ultimately to the long curving finger of the river's tidal estuary west of Lancaster. Apart from Lancaster, which the Lune impressively divides into two, the only town of any size on its route is Kirkby Lonsdale. However, there are many attractive small villages to be discovered, for example Newbiggin on Lune, Kelleth, Middleton, Rigmaden Park and Barbon, the last named famous among car enthusiasts for its hill climbs. The Lune has a number of tributaries, most of which flow directly down from the hills to the east but one, the **Greta**, is particularly memorable. This river rises in Whernside and then flows westward through Ingleton to the Lune, its course dramatically marked by waterfalls, gorges and caves, into which it sometimes vanishes completely.

The Lancaster Canal

The **Lancaster Canal** was first planned in 1771 and it is one of the few major man-made waterways in the north-west of England. It was not finally completed until early in the nineteenth century when the main line between Preston and Kendal was opened. Isolated from the main canal network, the Lancaster has always had a character of its own, with distinctive bridges, elegant aqueducts, and a predominantly rural 57 mile route through pleasant countryside. When it was first built, there was a tramway link between Preston Basin and the Leeds and Liverpool Canal, five miles to the south, but this did not last long. Commercial traffic on the Lancaster Canal continued until the 1940s and then it quietly declined. In 1955 the upper section between Tewitfield and Kendal was abandoned, but the lower section of over 42 miles is still open to pleasure boats. Despite its rural nature, the Lancaster Canal has many features of interest. Preston itself is not too exciting, but within a few miles the canal becomes entirely rural, passing old stables, mills and lift bridges. At Garstang there is an attractive basin, while at Galgate there is a short branch canal leading to Glasson with its old-fashioned docks and the sealock into the Lune estuary. North of Lancaster is a most impressive stone aqueduct, which is 640 feet long, over the river Lune, after which the canal passes Hest Bank, coming within a quarer of a mile of the sea.

After Carnforth, the home of Steamtown, one of the largest steam railway centres in England, the canal comes to its present rather abrupt terminus, at the foot of the M6 motorway. Ironically, the most attractive part of the canal is the closed section to Kendal, where there are locks, a large number of fine bridges and aqueducts, and a short tunnel at Hindcaster, as well as a more interesting landscape. Although parts of this section still hold water, the best way to explore the canal is on foot, for the towpath is a right of way throughout its length.

The Ribble and its tributaries

The **Ribble**, Lancashire's largest river, actually rises in North Yorkshire, high in the Pennines. Its course through the Yorkshire Dales is marked by tiny villages, notably New Houses, Horton in Ribblesdale, Helwith Bridge and Stackhouse until it reaches Settle. The landscape is dramatic through Ribblesdale but gradually the character of the river changes as it approaches the fertile valley that carries it towards Preston and its estuary. The ruins of Sawley Abbey are followed by Clitheroe, with its Norman castle and museum, and the Roman fort at Ribchester, which serves as a reminder of a far earlier period of civilisation when Lancashire was a wild and undeveloped region. Just outside Preston the river passes Samlesbury Hall, a fourteenth-century manor house, and then industry takes over as the Ribble skirts the southern border of Preston. Still a port of some significance, Preston has exploited the tidal estuary of the Ribble since Roman times. This long natural expanse of water connects the town with the sea, several miles to the west. The Ribble has a number of tributaries, the most attractive of which flow down from the fells west of Ribblesdale. These include the **Hodder**, which rises above Stocks Reservoir and then passes remote villages hidden high in the Forest of Bowland on its way to join the Ribble near Great Mitton. There are waterfalls and a packhorse bridge. Nearby, but flowing from the south is the **Calder**. This rises in the moors south of Burnley, near Claviger Gorge, passes through Burnley itself, and thence to Cawthorpe Hall and Great Harwood Abbey and Gardens. Not strictly a tributary, but sharing the same estuary as the Ribble is the river **Douglas**. Not much to look at today, this was an important waterway in the eighteenth century when it was made navigable from its estuary to Wigan. The navigation fell into disuse after the building of the Leeds and Liverpool Canal, which duplicated much of its route, from Wigan to Burscough Bridge, and from Burscough to Tarleton via the Rufford branch. Today only the last few miles are still navigable, linking Tarleton with the estuary, and connecting the Leeds and Liverpool Canal with the sea.

This elegant eighteenth-century bridge over the Lancaster Canal near Galgate, is one of several fine bridges that cross the canal.

The marina at Galgate on the Lancaster Canal, which is linked to Glasson and the Lune estuary by an eight-mile branch canal.

The Trans-Pennine Canals

One of the greatest challenges facing the canal builders of the eighteenth century was the need to link the industrial towns of South Yorkshire with Manchester and Liverpool. Colossal feats of engineering were required, for the Pennines stood in the way, and the builders themselves faced extraordinary hardships and yet, by the early years of the nineteenth century, no less than three trans-Pennine canals were in operation. The first of these was the **Rochdale Canal**, opened in 1804 to link the Bridgewater Canal in Manchester with the Calder and Hebble Navigation at Sowerby Bridge. The second was the **Huddersfield Narrow Canal**, opened in 1811 to join the Ashton Canal in Manchester with Huddersfield. The third was the **Leeds and Liverpool Canal**, built between 1770 and 1816. Today only one of these canals, the Leeds

and Liverpool, is open to traffic but the routes of the Rochdale and Huddersfield Narrow can still be traced and many of their most important and impressive features are still intact. Anyone wishing to understand the importance of canals in the eighteenth and early nineteenth centuries should explore these trans-Pennine routes.

The Rochdale Canal, with 92 locks in 33 miles, is a dramatic piece of engineering. It starts from the heart of Manchester, passing beneath Piccadilly and other familiar landmarks. The first nine locks are still in use, as this part of the Rochdale operates as a vital link between the Ashton and Peak Forest Canals, but the remainder is impassible. Between Manchester and Rochdale the surroundings are gloomy and industrial but full of history. After Rochdale the scenery improves, as the canal passes through Littleborough and Todmorden, climbing

all the time. After reaching its summit, the canal descends the dramatic Calder valley, through Hebden Bridge to its terminus basin in Sowerby Bridge. Traffic continued to use the Rochdale Canal until it was finally closed in 1952.

The Huddersfield Narrow Canal is equally dramatic, with 74 locks in under 20 miles, and at Standedge is the longest canal tunnel ever built in England. This astonishing structure, over 5600 yards long, was cut through solid rock and must have been a nightmare for the early Victorian boatmen, for its passage took over 3 hours. The route of the Huddersfield Narrow is partly industrial and partly through impressive Pennine scenery and, although parts have been filled in, it can easily be followed. Despite the ambitions of its builders, the Huddersfield Narrow Canal was never really successful, and regular through traffic had largely ceased by about 1904, although it was not

finally closed until 1944. There are plans to restore and reopen both these canals and work has begun, but it will be many years before boats can once again pass through Standedge tunnel. In the meantime, both can be explored on foot and the towpath is readily accessible by road in many places. To explore these canals is to understand the impact of the Industrial Revolution on the predominantly agricultural economy of England in the late eighteenth century.

The third route, the Leeds and Liverpool, is not only still open but also enjoys the most dramatic Pennine scenery. The canal is 127 miles long with 91 locks, and commercial traffic was active until the 1960s. The architecture of the canal itself and its surrounding buildings is impressive, with locks, bridges and mills all built from the local sandstone, hard, dark and impenetrable. There is a tunnel at Foulridge and many of the locks are grouped into

A fisherman on the Leeds and Liverpool Canal near West Marton. Canal fishing is very popular and many stretches are leased to fishing clubs.

The dramatic staircase of five locks at Bingley, one of the most exciting features of the Leeds and Liverpool Canal.

flights, with 23 close together at Wigan, and at Bingley the famous 5 locks staircase, one of the wonders of the waterways. The route of the canal is from the river Mersey by Liverpool docks, through Liverpool and Aintree to Wigan, then through Blackburn, Burnley, Gargrave, Skipton, Keighley, Saltaire to Leeds and the junction with the Aire and Calder Navigation. Branches also connect the Leeds and Liverpool to the Bridgewater Canal at Leigh, and to the sea via the Ribble estuary at Rufford. The scenery is wild and remote, a mixture of spectacular Pennine beauty. Victorian industry mingles with urban development and there are plenty of features of interest that have little to do with the canal itself, for example Kirkstall Abbey and Rufford Old Hall, besides all the surrounding towns and villages of the Yorkshire Dales.

The canals and rivers of Manchester
Although rarely recognised as such today, Manchester was once the centre of a great network of waterways that linked the city with many other parts of England. Some of these waterways date back to the Middle Ages, and the growth of Liverpool as a major trading port. Heavily industrialised since the eighteenth century, Manchester and its surrounding towns have inevitably swallowed up the wild and undeveloped countryside still so abundant further north. As part of the same process, the rivers and streams have been turned into commercial waterways, sources of power or basic drains. Clearly this is not an area of unspoilt countryside and gently flowing streams, yet there is plenty to see, for in many ways the Manchester region is a mirror of English history over the last two centuries.

At the heart of the region and the original cause of its prosperity, are two great rivers, the **Mersey** and the **Irwell**. Since the Middle Ages the Mersey has carried boats as far south as Warrington, and long before then Liverpool had begun to develop as a seaport. The long curving estuary of the Mersey provides a natural deep water and relatively sheltered harbour, and Liverpool and Birkenhead have been associated with ships and the sea for centuries. The growth of Manchester in the early eighteenth century created the need for adequate transport links with the port of Liverpool and this led to the development of the waterways. Manchester expanded around the river Irwell, a tributary of the Mersey and so the first step was to turn this into a navigation. The **Mersey and Irwell Navigation** was duly opened in 1736, and throughout the eighteenth century, as trade increased and the rival waterways came into being, it was steadily improved. It remained Manchester's main trade link with Liverpool and the sea until 1894 when the Manchester Ship Canal was opened. The first major rival to the Irwell Navigation was the **Bridgewater Canal**, part of which was opened in 1761. This canal, built to link Manchester with the Mersey at Runcorn, was the first major artificial waterway in the modern sense to be built in Britain and it effectively launched the canal age. The canal's instigator, the third Duke of Bridgewater, originally planned the canal, which was designed by the engineer James Brindley, to transport coal from his mines at Worsley Delph. As a commercial venture it

A view of the Leeds and Liverpool Canal passing through Saltaire, Sir Titus Salt's splendid Victorian industrial estate.

The Barton Aqueduct. This remarkable bridge swings on its central pivot, still full of water, to allow the passage of larger vessels on the Manchester Ship Canal.

was immediately successful and it was steadily expanded to connect with other canals, the Trent and Mersey at Preston Brook, the Leeds and Liverpool at Leigh and the Manchester Ship Canal. It continued to carry commercial traffic into the 1970s. Today the canal is largely used by pleasure boats but it is an interesting canal to explore, partly because of its associations with the start of the Industrial Revolution, and partly because the Bridgewater and its connections offer an unusual way to see the Manchester region. It also has two features of remarkable interest. The first of these is the Barton Aqueduct. When the canal was built, a brick aqueduct carried it over the river Irwell. This was demolished in the 1890s during the construction of the Manchester Ship Canal and replaced by the Barton Aqueduct. This unique and extraordinary device is a huge steel aqueduct that swings on a central pivot to allow ships to pass. To see this great machine in operation is to understand the ingenuity of the Victorian mind. The second remarkable feature is the original basin at Worsley where the Duke of Bridgewater shipped his coal to Manchester and elsewhere. Here, from a basin dramatically cut out of solid rock, are the entrances to a series of tunnels that linked the coal mines with the canal. This was a complete underground network, with locks, inclined planes and over 40 miles of waterway connecting the various coal mines and seams. Special narrow boats were built for this network and some can still be seen in the basin.

There are a number of other canals in the Manchester region that are now derelict but the enthusiast will find plenty to see. One of the earliest canals of the eighteenth century was built to connect St Helen's with the Mersey. The **St Helen's Canal** was not finally closed until the 1940s and much remains, including locks at Widnes and St Helen's and workshops and warehouses at Winwick. However, the canal is rapidly disappearing under various development schemes. More substantial but less easy to explore, requiring foot work rather than a car, is the **Manchester, Bolton and Bury Canal**. Opened in 1797, this canal was built to serve the coal trade and continued to trade commercially until the 1930s. It also operated a successful passenger service during its early years. Today, its route can still be traced. There are locks to be seen near Salford and Prestolee, aqueducts over the river Irwell (the best of these is also at Prestolee) and the towpath provides an interesting walk into Bury.

The final waterway, and obviously the most important, is the **Manchester Ship Canal** itself. Opened in 1894, this major undertaking turned Manchester from a canal city into an international port. Ships up to 600 feet long, 65 feet wide and drawing 28 feet of water can still sail into the heart of the city, 36 miles from the sea. Although part of its route followed the river Mersey and Irwell, much of the Ship Canal was dug from scratch. There are five

The Manchester Ship Canal under construction. The largest man-made waterway in England, it is also the first to have been built with mechanical equipment.

The unexpected sight of a coaster sailing through the fields miles from the sea, one of the more dramatic pleasures of the Manchester Ship Canal.

huge locks and a number of bridges, which either swing to allow the passage of ships or, like the M6 motorway bridge, cross the canal at high level. In some ways the canal itself is a linear dock, lined with quays, jetties and terminals from its connection with the Mersey at Eastham to Manchester docks. The Manchester Ship Canal is the only waterway in England to have been built on this scale and, although traffic today is gradually diminishing, it is still worth a visit, for its engineering, and for the spectacle of huge ships sailing through the countryside west of Manchester.

ACKNOWLEDGMENTS

Illustrations have been reproduced by kind permission of the following:

Aerofilms: 25, 41, 43, 47

John Bethell: 9, 11, 56–7, 59, 61, 82, 84, 98, 101, 102, 105, 113, 124

Janet and Colin Bord: 6, 12, 20, 31, 33, 34–5, 37 bottom, 58, 60, 64, 70, 73 right, 93, 99, 119, right, 122, 131, 135 top

British Museum: 104

British Waterways Board: 76 bottom left, 86, 87, 90, 95, 96, 139, 143

Fotobank: endpapers, 134–5

Gainsborough Library, Lincolnshire: 10 bottom

Angelo Hornak: 29

Lady Lever Art Gallery: 46

Manchester Ship Canal Company: 145

Mansell Collection: 10 top, 119 left

Sheila and Oliver Mathews: copyright page, 28, 35 right, 40 bottom, 126

National Trust: 19, 68

Derek Pratt: 16, 18, 24 top, 27 top, 32 top, 42, 50, 52, 53 left, 54 top, 55, 69, 71, 74, 75, 76 top left, 77, 79, 81, 88–9, 91, 92, 94, 109, 115, 120, 125, 129, 141, 142, 144, 146

Arthur Oglesby: 36

Royal Academy: 103

Brian and Sally Shuel: 17 bottom, 24 bottom, 54 bottom, 111, 121 right, 123, 127, 138

Edwin Smith: 72

Swanston Graphics: 14–15, 30–1, 44–5, 62, 76 top right, 80, 100, 118, 132

Patrick Thurston: frontispiece, 22, 48 bottom, 85, 108, 117, 136, 137

Michael Turner/The Sunday Times Magazine: 7

Roy Westlake: 8, 17 top, 21, 26, 53 right, 66, 73 left, 97, 107

Derek Widdicombe: 38, 106, 110, 128, 130

Trevor Wood: 23, 48 top, 83, 112

Wookey Hole: 27 bottom

Weidenfeld and Nicolson Archives: 65, 67

INDEX